Digital Signal Processing

Everything you need to know to get started

Michael Parker
Altera Corporation

AMSTERDAM • BOSTON • HEIDELBERG • LONDON
NEW YORK • OXFORD • PARIS • SAN DIEGO
SAN FRANCISCO • SINGAPORE • SYDNEY • TOKYO
Newnes is an imprint of Elsevier

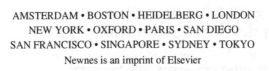

ELSEVIER

Newnes

Newnes is an imprint of Elsevier
30 Corporate Drive, Suite 400, Burlington, MA 01803, USA
The Boulevard, Langford Lane, Kidlington, Oxford OX5 1GB, UK

Library of Congress Cataloging-in-Publication Data
Parker, Michael, 1963-
   Digital signal processing 101: everything you need to know to get started / Michael Parker.
      p. cm.
   Includes bibliographical references and index.
   ISBN 978-1-85617-921-8 (alk. paper)
   1. Signal processing–Digital techniques. I. Title.
   TK5102.9.P385 2010
   621.382′2–dc22   2010002121

British Library Cataloguing-in-Publication Data
A catalogue record for this book is available from the British Library.

For information on all Newnes publications
visit our Web site at www.elsevierdirect.com

10 11 12 13 14   10 9 8 7 6 5 4 3 2 1
Printed in the United States of America

Table of Contents

Introduction . . . . *viii*
Acknowledgments . . . . *x*

Chapter 1: Numerical Representation *1*

   1.1 Integer Fixed-Point Representation 2
   1.2 Fractional Fixed-Point Representation 4
   1.3 Floating-Point Representation 7

Chapter 2: Complex Numbers and Exponentials *9*

   2.1 Complex Addition and Subtraction 9
   2.2 Complex Multiplication 10
   2.3 Complex Conjugate 15
   2.4 The Complex Exponential 16
   2.5 Measuring Angles in Radians 18

Chapter 3: Sampling, Aliasing, and Quantization *21*

   3.1 Nyquist Sampling Rule 25
   3.2 Quantization 27

Chapter 4: Frequency Response *31*

   4.1 Frequency Response and the Complex Exponential 31
   4.2 Normalizing Frequency Response 33
   4.3 Sweeping across the Frequency Response 34
   4.4 Example Frequency Responses 35
   4.5 Linear Phase Response 37
   4.6 Normalized Frequency Response Plots 38

Chapter 5: Finite Impulse Response (FIR) Filters *41*

   5.1 FIR Filter Construction 41
   5.2 Computing Frequency Response 45
   5.3 Computing Filter Coefficients 49
   5.4 Effect of Number of Taps on Filter Response 52

Chapter 6: Windowing *57*

   6.1 Truncation of Coefficients 57
   6.2 Tapering of Coefficients 58
   6.3 Example Coefficient Windows 59

Chapter 7: Decimation and Interpolation **63**
   7.1 Decimation 63
   7.2 Interpolation 67
   7.3 Resampling by Non-Integer Value 70

Chapter 8: Infinite Impulse Response (IIR) Filters **73**
   8.1 IIR and FIR Filter Characteristic Comparison 74
   8.2 Bilinear Transform 76
   8.3 Frequency Prewarping 78

Chapter 9: Complex Modulation and Demodulation **81**
   9.1 Modulation Constellations 81
   9.2 Modulated Signal Bandwidth 84
   9.3 Pulse-Shaping Filter 85
   9.4 Raised Cosine Filter 88

Chapter 10: Discrete and Fast Fourier Transforms (DFT, FFT) **97**
   10.1 DFT and IDFT Equations 98
   10.2 Fast Fourier Transform (FFT) 106
   10.3 Filtering Using the FFT and IFFT 110
   10.4 Bit Growth in FFTs 111
   10.5 Bit-Reversal Addressing 112

Chapter 11: Digital Upconversion and Downconversion **113**
   11.1 Digital Upconversion 114
   11.2 Digital Downconversion 117
   11.3 IF Subsampling 118

Chapter 12: Error Correction Coding **125**
   12.1 Linear Block Encoding 126
   12.2 Linear Block Decoding 127
   12.3 Minimum Coding Distance 130
   12.4 Convolutional Encoding 131
   12.5 Viterbi Decoding 134
   12.6 Soft Decision Decoding 140
   12.7 Cyclic Redundancy Check 141
   12.8 Shannon Capacity and Limit Theorems 142

Chapter 13: Analog and TDMA Wireless Communications **143**
   13.1 Early Digital Innovations 144
   13.2 Frequency Modulation 145
   13.3 Digital Signal Processor 146
   13.4 Digital Voice Phone Systems 147
   13.5 TDMA Modulation and Demodulation 148

Chapter 14: CDMA Wireless Communications *151*

14.1   Spread Spectrum Technology 151
14.2   Direct Sequence Spread Spectrum 152
14.3   Walsh Codes 153
14.4   Concept of CDMA 155
14.5   Walsh Code Demodulation 155
14.6   Network Synchronization 158
14.7   RAKE Receiver 159
14.8   Pilot PN Codes 159
14.9   CDMA Transmit Architecture 160
14.10 Variable Rate Vocoder 162
14.11 Soft Handoff 163
14.12 Uplink Modulation 163
14.13 Power Control 164
14.14 Higher Data Rates 166
14.15 Spectral Efficiency Considerations 166
14.16 Other CDMA Technologies 167

Chapter 15: OFDMA Wireless Communications *169*

15.1   WiMax and LTE 169
15.2   OFDMA Advantages 170
15.3   Orthogonality of Periodic Signals 171
15.4   Frequency Spectrum of Orthogonal Subcarrier 173
15.5   OFDM Modulation 175
15.6   Intersymbol Interference and the Cyclic Prefix 177
15.7   MIMO Equalization 180
15.8   OFDMA System Considerations 181
15.9   OFDMA Spectral Efficiency 182
15.10 OFDMA Doppler Frequency Shift 183
15.11 Peak to Average Ratio 183
15.12 Crest Factor Reduction 185
15.13 Digital Predistortion 188
15.14 Remote Radio Head 189

Chapter 16: Radar Basics *191*

16.1   Radar Frequency Bands 191
16.2   Radar Antennas 192
16.3   Radar Range Equation 195
16.4   Stealth Aircraft 196
16.5   Pulsed Radar Operation 196
16.6   Pulse Compression 197
16.7   Pulse Repetition Frequency 197
16.8   Detection Processing 200

Chapter 17: Pulse Doppler Radar *201*

    17.1 Doppler Effect 201
    17.2 Pulsed Frequency Spectrum 203
    17.3 Doppler Ambiguities 205
    17.4 Radar Clutter 206
    17.5 PRF Trade-offs 208
    17.6 Target Tracking 210

Chapter 18: Synthetic Array Radar *213*

    18.1 SAR Resolution 213
    18.2 Pulse Compression 214
    18.3 Azimuth Resolution 214
    18.4 SAR Processing 218
    18.5 SAR Doppler Processing 219
    18.6 SAR Impairments 221

Chapter 19: Introduction to Video Processing *223*

    19.1 Color Spaces 223
    19.2 Interlacing 225
    19.3 Deinterlacing 225
    19.4 Image Resolution and Bandwidth 227
    19.5 Chroma Scaling 228
    19.6 Image Scaling and Cropping 228
    19.7 Alpha Blending and Compositing 229
    19.8 Video Compression 229
    19.9 Video Interfaces 230

Chapter 20: Implementation Using Digital Signal Processors *233*

    20.1 DSP Processor Architectural Enhancements 233
    20.2 Scalability 238
    20.3 Floating Point 238
    20.4 Design Methodology 239
    20.5 Managing Resources 239
    20.6 Ecosystem 240

Chapter 21: Implementation Using FPGAs *243*

    21.1 FPGA Design Methodology 244
    21.2 DSP Processor or FPGA Choice 245
    21.3 Design Methodology Considerations 246
    21.4 Dedicated DSP Circuit Blocks in FPGAs 247
    21.5 Floating Point in FPGAs 253
    21.6 Ecosystem 254
    21.7 Future Trends 255

Appendix A: Q Format Shift with Fractional Multiplication *257*
Appendix B: Evaluation of FIR Design Error Minimization *259*
Appendix C: Laplace Transform *263*
Appendix D: Z-Transform *267*
Appendix E: Binary Field Arithmetic *271*

Index *273*

Introduction

This book is intended for those who work in or provide components for industries that are made possible by digital signal processing, or DSP. Sample industries are wireless mobile phone and infrastructure equipment, broadcast and cable video, DSL modems, satellite communications, medical imaging, audio, radar, sonar, surveillance, electrical motor control—this list goes on. While the engineers who implement these systems must be very familiar with DSP, there are many others—executive and midlevel management, marketing, technical sales and field engineers, business development, and others—who can benefit from a basic knowledge of the fundamental principles of DSP.

Others who are a potential audience include those interested in studying or working in any of these areas. High school seniors or undeclared college majors considering a future in the industries made possible by DSP technology may gain sufficient understanding that enables them to decide whether to continue further.

That, then, is the purpose of this book: to provide a basic tutorial on DSP. This topic seems to have a dearth of easy-to-read and understand explanations. Unlike most technical resources, this is a treatment in which mathematics is minimized and intuitive understanding maximized. This book attempts to explain many difficult concepts like sampling, aliasing, imaginary numbers, and frequency response using easy-to-understand examples. In addition, there is an overview of the DSP functions and implementation used in several DSP-intensive fields or applications, from error correction to CDMA mobile communication to airborne radar systems.

So this book is intended for those of you who, like me, are somewhat dismayed when presented with a blackboard (or whiteboard) full of equations as an explanation on DSP.

The intended readers include those who have absolutely no previous experience with DSP, but are comfortable with high-school-level math skills. Many technical details have been deliberately left out, in order to emphasize just the key concepts. While this book is not expected to be used as a university-course-level text, it can initiate readers prior to tackling a proper text on DSP. But it may also be all you need to talk intelligently to other people involved in a DSP-centric industry and understand many of the fundamental concepts.

To start with, just what is DSP? Well, DSP is performing operations on a digital signal of some sort and using a digital semiconductor device. Most commonly, multipliers and adders are used. If you can multiply and add, you can probably understand DSP. Actually, signal processing was around long before digital electronics. Examples of this are radios and TVs. Early tuners used analog circuits with variable capacitors to dial a station. Resistors, capacitors, and vacuum tubes were used to either attenuate or amplify different frequencies or to provide frequency shifting. These are examples of basic signal processing applications. The signals were analog signals, and the circuits doing the processing were analog, as was the final output.

Today, most signal processing is performed digitally. The reason is that digital circuits have progressively become cheaper and faster, as well as due to the inherent advantages of repeatability, tolerance, and consistency that digital circuits enjoy compared to analog circuits.

If the signal is not in a digital form, then it must first by converted, or digitized. A device called an analog-to-digital converter (ADC) is used. If the output signal needs to be analog, then it is converted back using a digital-to-analog converter (DAC). Of course, many signals are already digitized and can be processed by digital circuits directly.

DSP is at the heart of a wide range of everyday devices in our lives, although many people are unaware of this. A few everyday examples are cellular phones, DSL modems, digital hearing aids, MRI and ultrasound equipment, audio equipment, set top boxes, flat-screen televisions, satellite communications, and DVD players.

As promised, the mathematics will be minimized, but it cannot be eliminated altogether. Some basic trigonometry and the use of complex numbers are unavoidable, so an early chapter is included to introduce these concepts, using as simple examples as possible. There is also one appendix section where very basic calculus is used, but this is not essential to the overall understanding.

Acknowledgments

This book grew out of a need for Altera marketing and technical sales people to have an intuitive-level understanding of DSP fundamentals and applications, in order to better work on issues that our customers face as they implement DSP systems. I am grateful to the Altera management for the support this book has received, in particular from Steve Mensor and Chris Balough.

My understanding of the topics in this book is based on many years of engineering implementation work and collaboration and explanations from many of my colleagues at multiple firms over the years. More recently, within Altera, many people have contributed to my knowledge in these areas. I would like to especially acknowledge a few people who have been helpful both in DSP domain and relevant applications and implementations. Within Altera engineering, this includes Volker Mauer, Martin Langhammer, and Mike Fitton. Within the Altera technical sales organization, people who have been especially helpful to my understanding of some of the relevant DSP applications include Colman Cheung, Ben Esposito, Brian Kurtz, and Mark Santoro.

Within Altera publications, James Adams has been instrumental in getting this project off the ground and working with the publisher.

Finally, the support of my wife, Zaida, and daughter, Ariel, have been most important. This book has been primarily an "evenings and weekends" project, and their patience has been essential.

CHAPTER 1

Numerical Representation

To process a signal digitally, it must be represented in a digital format. This point may seem obvious, but it turns out that there are a number of different ways to represent numbers, and this representation can greatly affect both the result and the number of circuit resources required to perform a given operation. This chapter is focused more for people who are implementing digital signal processing (DSP) and is not really required to understand DSP fundamental concepts.

Digital electronics operate on bits, of course, which are used to form binary words. The bits can be represented as binary, decimal, octal, hexadecimal, or another form. These binary numbers can be used to represent "real" numbers. There are two basic types of arithmetic used in DSP: floating point and fixed point. Fixed-point numbers have a fixed decimal point as part of the number. Examples are 1234 (the same as 1234.0), 12.34, and 0.1234. This is the type of number we normally use every day. A floating-point number has an exponent. The most common example is scientific notation, used on many calculators. In floating point, 1,200,000 would be expressed as 1.2×10^6, and 0.0000234 would be expressed as 2.34×10^{-5}. Most of our discussion will focus on fixed-point numbers, as they are most commonly found in DSP applications. Once we understand DSP arithmetic issues with fixed-point numbers, then there is short discussion of floating-point numbers.

In DSP, we pretty much exclusively use signed numbers, meaning that there are both positive and negative numbers. This leads to the next point, which is how to represent the negative numbers.

In signed fixed-point arithmetic, the binary number representations include a sign, a radix or decimal point, and the magnitude. The sign indicates whether the number is positive or negative, and the radix (also called decimal) point separates the integer and fractional parts of the number.

The sign is normally determined by the leftmost, or most significant bit (MSB). The convention is that a zero is used for positive and one for negative. There are several formats to represent negative numbers, but the almost universal method is known as *2s complement*. This is the method discussed here.

Furthermore, fixed-point numbers are usually represented as either integer or fractional. In integer representation, the decimal point is to the right of the least significant bit (LSB), and there is no fractional part in the number. For an 8-bit number, the range that can be represented is from -128 to $+127$, with increments of 1.

Digital Signal Processing 101. DOI: 10.1016/B978-1-85617-921-8.00005-5

In fractional representation, the decimal point is often just to the right of the MSB, which is also the sign bit. For an 8-bit number, the range that can be represented is from -1 to $+127/128$ (almost $+1$), with increments of $1/128$. This may seem a bit strange, but in practice, fractional representation has advantages, as will be explained.

This chapter presents several tables, with each row giving equivalent binary and hexadecimal numbers. The far right column gives the actual value in the chosen representation—for example, 16-bit integer representation. The actual value represented by the hex/binary numbers depends on which representation format is chosen.

1.1 Integer Fixed-Point Representation

The following table provides some examples showing the 2s complement integer fixed-point representation.

Table 1.1: 8-Bit integer representation

Binary	Hexadecimal	Actual Decimal Value
0111 1111	0x7F	127
0111 1110	0x7E	126
0000 0010	0x02	2
0000 0001	0x01	1
0000 0000	0x00	0
1111 1111	0xFF	-1
1111 1110	0xFE	-2
1000 0001	0x81	-127
1000 0000	0x80	-128

The 2s complement representation of the negative numbers may seem nonintuitive, but it has several very nice features. There is only one representation of 0 (all 0s), unlike other formats that have a "positive" and "negative" zero. Also, addition and multiplication of positive and negative 2s complement numbers work properly with traditional digital adder and multiplier structures. A 2s complement number range can be extended to the left by simply replicating the MSB (sign bit) as needed, without changing the value.

The way to interpret a 2s complement number is to use the mapping for each bit shown in the following table. A 0 bit in a given location of the binary word means no weight for that bit. A 1 in a given location means to use the weight indicated. Notice the weights double with each bit moving left, and the MSB is the only bit with a negative weight. You should satisfy yourself that all negative numbers will have an MSB of 1, and all positive numbers and zero have an MSB of 0.

Table 1.2: 2s complement bit weighting with 8 bit words

8-Bit Signed Integer	MSB							LSB
Bit weight	−128	64	32	16	8	4	2	1
Weight in powers of 2	-2^7	2^6	2^5	2^4	2^3	2^2	2^1	2^0

This can be extended to numbers with larger number of bits. Following is an example with 16 bits. Notice how the numbers represented in a lesser number of bits (e.g., 8 bits) can be easily put into 16-bit representation by simply replicating the MSB of the 8-bit number eight times and tacking onto the left to form a 16-bit number. Similarly, as long as the number represented in the 16-bit representation is small enough to be represented in 8 bits, the leftmost bits can simply be shaved off to move to the 8-bit representation. In both cases, the decimal point stays to the right of the LSB and does not change location. This can be seen easily by comparing, for example, the representation of −2 in the 8-bit representation table and again in the 16-bit representation table.

Table 1.3: 16-Bit signed integer representation

Binary	Hexadecimal	Actual Decimal Value
0111 1111 1111 1111	0×7FFF	32,767
0111 1111 1111 1110	0×7FFE	32,766
0000 0000 1000 0000	0×0080	128
0000 0000 0111 1111	0×007F	127
0000 0000 0111 1110	0×007E	126
0000 0000 0000 0010	0×0002	2
0000 0000 0000 0001	0×0001	1
0000 0000 0000 0000	0×0000	0
1111 1111 1111 1111	0×FFFF	−1
1111 1111 1111 1110	0×FFFE	−2
1111 1111 1000 0001	0×FF81	−127
1111 1111 1000 0000	0×FF80	−128
1111 1111 0111 1111	0×FF80	−129
1000 0000 0000 0001	0×FF80	−32,767
1000 0000 0000 0000	0×FF80	−32,768

Table 1.4: 2s complement bit weighting with 16 bit word

MSB															LSB
−32, 768	16, 384	8192	4096	2048	1024	512	256	128	64	32	16	8	4	2	1
-2^{15}	2^{14}	2^{13}	2^{12}	2^{11}	2^{10}	2^9	2^8	2^7	2^6	2^5	2^4	2^3	2^2	2^1	2^0

Now, let's look at some examples of trying to adding combinations of positive and negative 8-bit numbers together using a traditional unsigned digital adder. We throw away the carry bit from the last (MSB) adder stage.

Case #1: Positive and negative number sum
+15	0000 1111	0x0F
−1	1111 1111	0xFF
+14	0000 1110	0x0E

Case #2: Positive and negative number sum
−31	1110 0001	0xE1
+16	0001 0000	0x80
−15	1111 0000	0xF0

Case #3: Two negative numbers being summed
−31	1110 0001	0xE1
−64	1100 0000	0xC0
−95	1010 0001	0xF0

Case #4: Two positive numbers being summed; result exceeds range
+64	0100 0000	0x40
+64	0100 0000	0x40
+128	1000 0000**	0x80**

**Notice all the results are correct, except the last case. The reason is that the result, +128, cannot be represented in the range of an 8-bit 2s complement number.

Integer representation is often used in many software applications because it is familiar and works well. However, in DSP, integer representation has a major drawback. In DSP, there is a lot of multiplication. When you multiply a bunch of integers together, the results start to grow rapidly. It quickly gets out of hand and exceeds the range of values that can be represented. As we saw previously, 2s complement arithmetic works well, as long as you do not exceed the numerical range. This has led to the use of fractional fixed-point representation.

1.2 Fractional Fixed-Point Representation

The basic idea behind fractional fixed-point representation is all values are in the range from +1 to −1, so if they are multiplied, the result will not exceed this range. Notice that, if you want to convert from integer to 8-bit signed fractional, the actual values are all divided by 128. This maps the integer range of +127 to −128 to almost +1 to −1.

Table 1.5: 8-Bit fractional representation

Binary	Hexadecimal	Actual Decimal Value
0111 1111	0x7F	127/128 = 0.99219
0111 1110	0x7E	126/128 = 0.98438
0000 0010	0x02	2/128 = 0.01563
0000 0001	0x01	1/128 = 0.00781
0000 0000	0x00	0
1111 1111	0xFF	−1/128 = −0.00781
1111 1110	0xFE	−2/128 = −0.01563
1000 0001	0x81	−127/128 = −0.99219
1000 0000	0x80	−1.00

Table 1.6: 2s complement weighting for 8 bit fractional word

8-Bit Signed Fractional	MSB							LSB
Weight	−1	1/2	1/4	1/8	1/16	1/32	1/64	1/128
Weight in powers of 2	-2^0	2^{-1}	2^{-2}	2^{-3}	2^{-4}	2^{-5}	2^{-6}	2^{-7}

Table 1.7: 16-Bit signed fractional representation

Binary	Hexadecimal	Actual Decimal Value
0111 1111 1111 1111	0x7FFF	32,767/32,768
0111 1111 1111 1110	0x7FFE	32,766/32,768
0000 0000 1000 0000	0x0080	128/32,768
0000 0000 0111 1111	0x007F	127/32,768
0000 0000 0111 1110	0x007E	126/32,768
0000 0000 0000 0010	0x0002	2/32,768
0000 0000 0000 0001	0x0001	1/32,768
0000 0000 0000 0000	0x0000	0
1111 1111 1111 1111	0xFFFF	−1/32,768
1111 1111 1111 1110	0xFFFE	−2/32,768
1111 1111 1000 0001	0xFF81	−127/32,768
1111 1111 1000 0000	0xFF80	−128/32,768
1111 1111 0111 1111	0xFF7F	−129/32,768
1000 0000 0000 0001	0x8001	−32,767/32,768
1000 0000 0000 0000	0x8000	−1

Table 1.8: 2s complement weighting for 16 bit fractional word

MSB															LSB
−1	1/2	1/4	1/8	1/16	1/32	1/64	1/128	1/256	1/512	1/1024	1/2048	1/4096	1/8192	1/16,384	1/32,768
-2^0	2^{-1}	2^{-2}	2^{-3}	2^{-4}	2^{-5}	2^{-6}	2^{-7}	2^{-8}	2^{-9}	2^{-10}	2^{-11}	2^{-12}	2^{-13}	2^{-14}	2^{-15}

Fractional fixed point is often expressed in *Q format*. The representation shown above is Q15, which means that there are 15 bits to the right of the radix or decimal point. It might also be called Q1.15, meaning that there are 15 bits to right of the decimal point and one bit to the left.

The key property of fractional representation is that the numbers grow smaller, rather than larger, during multiplication. And in DSP, we commonly sum the results of many multiplication operations. In integer math, the results of multiplication can grow large quickly (see the following example). And when we sum many such results, the final sum can be very large, easily exceeding the ability to represent the number in a fixed-point integer format.

As an analogy, think about trying to display an analog signal on an oscilloscope. You need to select a voltage range (volts/division) in which the signal amplitude does not exceed the upper and lower range of the display. At the same time, you want the signal to occupy a reasonable part of the screen, so the detail of the signal is visible. If the signal amplitude occupies only 1/10 of a division, for example, it is difficult to see the signal.

To illustrate this situation, imagine using a 16-bit fixed point, and the signal has a value of 0.75 decimal or 24,676 in integer (which is 75% of full scale), and is multiplied by a coefficient of value 0.50 decimal or 16,384 integer (which is 50% of full scale).

$$
\begin{array}{ll}
0.75 & 24{,}576 \\
\times\ 0.50 & \times\ 16{,}384 \\
\hline
0.375 & 402{,}653{,}184
\end{array}
$$

Now the larger integer number can be shifted right after every such operation to scale the signal within a range that it can be represented, but most DSP designers prefer to represent numbers as fractional because it is a lot easier to keep track of the decimal point.

Now consider multiplication again. If two 16-bit numbers are multiplied, the result is a 32-bit number. As it turns out, if the two numbers being multiplied are Q15, you might expect the result in the 32-bit register to be a Q31 number (MSB to the left of the decimal point, all other bits to the right). Actually, the result is in Q30 format; the decimal point has shifted down to the right. Most DSP processors will automatically left shift the multiplier output to compensate for this when operating in fractional arithmetic mode. In Field Programmable Gate Array (FPGA) or hardware design, the designer may have to take this into account when connecting data buses between different blocks. Appendix A explains the need for this extra left shift in detail, as it will be important for those implementing fractional arithmetic on FPGAs or DSPs.

0×4000	value = ½ in Q15
× 0×2000	value = ¼ in Q15

0×0800 0000	value = 1/16 in Q31

After left shifting by one, we get

0×1000 0000	value = 1/8 in Q31—the correct result!

If we use only the top 16-bit word from multiplier output, after the left shift, we get

0×1000	value = 1/8 in Q15—again, the correct result!

1.3 Floating-Point Representation

Many of the complications encountered using the preceding methods can be avoided by using floating-point format. Floating-point format is basically like scientific notation on your calculator. Because of this, a floating-point number can have a much greater dynamic range than a fixed-point number with an equivalent number of bits. Dynamic range means the ability to represent very small numbers to very large numbers.

The floating-point number has both a mantissa (which includes sign) and an exponent. The mantissa is the value in the main field of your calculator, and the exponent is off to the side, usually as a superscript. Each is expressed as a fixed-point number (meaning the decimal point is fixed). The mantissa is usually in signed fractional format, and the exponent in signed integer format. The size of the mantissa and exponent in number of bits will vary depending on which floating-point format is used.

The following table shows two common 32-bit formats: "two word" and IEEE 754.

Table 1.9: Floating point format summary

Floating-Point Formats	No. of Mantissa Bits	No. of Exponent Bits
"two word"	16, in signed Q15	16, signed integer
IEEE_STD-754	23, in unsigned Q15, plus 1 bit to determine sign (not 2s complement!)	8, unsigned integer, biased by +127

To convert a fixed-point number in floating-point representation, we shift the fixed number left until there are no redundant sign bits. This process is called *normalization*. The number of these shifts determines the exponent value.

The drawback of floating-point calculations is the resources required. When adding or subtracting two floating-point numbers, we must first adjust the number with smaller absolute value so that its exponent is equal to the number with larger absolute value; then we can add the two mantissas. If the mantissa result requires a left or right shift to represent, the exponent is adjusted to account for this shift. When multiplying two floating-point numbers, we multiply the mantissas and then sum the exponents. Again, if the mantissa result requires a left or right shift to represent, the new exponent must be adjusted to account for this shift. All of this requires considerably more logic than fixed-point calculations and often must run at much lower speed (although recent advances in FPGA floating-point implementation may significantly narrow this gap). For this reason, most DSP algorithms use fixed-point arithmetic, despite the onus on the designer to keep track of where the decimal point is and ensure that the dynamic range of the signal never exceeds the fixed-point representation or else becomes so small that quantization error becomes insignificant. We will see more on quantization error in a later chapter.

If you are interested in more information on floating-point arithmetic, there are many texts that go into this topic in detail, or you can consult the IEEE_STD-754 document.

CHAPTER 2

Complex Numbers and Exponentials

Complex numbers are among those things many of us were taught a long time ago and have long since forgotten. Unfortunately, they are important in digital communications and DSP, so we need to resurrect them.

What we were taught—and some of us vaguely remember—is that a complex number has a "real" and an "imaginary" part, and the imaginary part is the square root of a negative number, which is really a nonexistent number. This explanation right away sounds fishy, and while it's technically true, there is a much more intuitive way of looking at it.

The whole reason for complex numbers is that we are going to need a two-dimensional number plane to understand DSP. The traditional number line extends from plus infinity to minus infinity, along a single line. To represent many of the concepts in DSP, we need two dimensions. This requires two orthogonal axes, like a North–South line and an East–West line. For the arithmetic to work out, one line, usually depicted as the horizontal line, is the real number line. The other vertical line is the imaginary line. All imaginary numbers are prefaced by "j," which is defined as the square root of –1. Don't get confused by this imaginary number stuff, but rather view "j" as an arbitrary construct we will use to differentiate the horizontal axis (normal) numbers from those on the vertical axis. This is the essence of this whole chapter.

As depicted in Figure 2.1, any complex number Z has a real and an imaginary part, and is expressed as $X + j \cdot Y$ or just $X + jY$. The value of X and Y for any point is determined by the distance one must travel in the direction of each axis to arrive at the point. It can also be visualized as a point on the complex number plane, or as a vector originating at the origin and terminating at the point. We need to be able to do arithmetic with complex numbers. There has to be a way to keep track of the vertical and horizontal components. That's where the "j" comes in (in some texts, "i" is used instead).

2.1 Complex Addition and Subtraction

Adding and subtracting are simple: just add and subtract the vertical and horizontal components separately. The "j" helps us keep from mixing the vertical and horizontal components. For example,

$$(3 + j4) - (1 - j6) = 2 + j10$$

Digital Signal Processing 101. DOI: 10.1016/B978-1-85617-921-8.00006-7

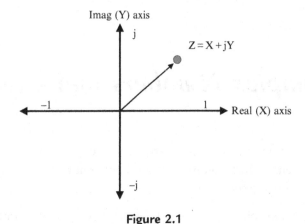

Figure 2.1

Scaling a complex number is simply scaling each component:

$$4 \cdot (3 + j4) = 12 + j16$$

2.2 Complex Multiplication

Multiplication gets a little trickier and is harder to visualize graphically. Here is the way the mechanics of it work:

$$(A + jB) \cdot (C + jD) = A \cdot C + jB \cdot C + A \cdot jD + jB \cdot jD = AC + jBC + jAD + j^2BD$$

Now remember that j^2 is, by definition, equal to –1. After collecting terms, we get

$$AC + jBC + jAD + j^2BD = AC + jBC + jAD - BD = (AC - BD) + j(BC + AD)$$

The result is another complex number, with $AC - BD$ being the real part and $BC + AD$ being the imaginary part (remember, *imaginary* just means the vertical axis, while *real* is the horizontal axis). This result is just another point on the complex plane.

The mechanics of this arithmetic may be simple, but we need to be able to visualize what is really happening. To do this, we need to introduce polar (R, Ω) representation. Until now, we have been using Cartesian (X, Y) coordinates, which means each location on the complex number plane is specified by the distance along each of the two axes (like longitude and latitude on the earth's surface). Polar representation replaces these two parameters, which can specify any point on the complex plane, with another set of two parameters, which also can specify any point on the complex plane. The two new parameters are the magnitude and angle. The *magnitude* is simply the length of the line or vector from the origin to the point. The *angle* is defined as the angle of this line starting at the positive X axis and arcing counterclockwise.

This relationship is shown in Figure 2.2, where the same point $Z = X + jY$ is identified by having radius R (length of the vector from origin to the point) with angle Ω specified counterclockwise from the positive real axis.

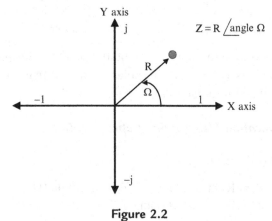

Figure 2.2

Any point Z on the graph may be specified as $X + jY$ or R with angle Ω.

To understand the relationships between these, we need to go back to basic high school math. Consider the right triangle formed in Figure 2.3, with sides X, Y, R, and angle Ω.

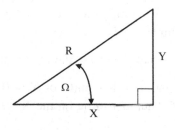

Figure 2.3

Remember the Pythagorean theorem? For any right triangle, $X^2 + Y^2 = R^2$.

Also, remember that sine is defined as the length of the opposite side divided by the hypotenuse, and cosine is defined as the length of the adjacent side divided by the hypotenuse (but it must be a right triangle). Tangent is defined as the opposite over the adjacent side. So we get the following relationships, which can be used to convert between Cartesian (X,Y) and polar (R, Ω).

2.2.1 Polar Representation

$$\begin{aligned}
\sin(\Omega) &= (Y/R) \quad \Rightarrow \quad Y = R \cdot \sin(\Omega) \\
\cos(\Omega) &= (X/R) \quad \Rightarrow \quad X = R \cdot \cos(\Omega) \\
X^2 + Y^2 &= R^2 \quad \Rightarrow \quad R = \text{sqrt}\,(X^2 + Y^2) \\
\tan(\Omega) &= (X/Y) \quad \Rightarrow \quad \Omega = \arctan\,(X/Y)
\end{aligned}$$

The reason for this little foray into polar representation is that multiplication (and division) of complex numbers is very easy in polar form, and that angles in the complex plane can be easily visualized in polar form.

2.2.2 Complex Multiplication Using Polar Representation

We will define two points, Z_1 and Z_2:

$$Z_1 = R_1 \text{ angle } (\Omega_1) \qquad Z_2 = R_2 \text{ angle } (\Omega_2)$$
$$Z_1 \cdot Z_2 = (R_1 \cdot R_2) \text{ angle } (\Omega_1 + \Omega_2)$$

This means that with any two complex numbers, the magnitude, or distance from the origin to the radius, gets multiplied together to form the new magnitude. This makes sense intuitively. The angles of the two complex numbers get added together to form the new angle. Not so intuitive, so let's try a few examples to get the hang of it.

Let's use real numbers to start; a real number is just a complex number with the "j" part equal to zero. Real numbers are "simplified" versions of complex numbers, so any arithmetic rules on a complex number had better work with the real numbers.

$$Z = R \text{ angle } (\Omega)\text{—for real numbers, the angle must be equal to } 0°$$

Then $X = R \cdot \cos(0)$ or $X = R$ and $Y = R \cdot \sin(0)$ or $Y = 0$. This is what we expect: the Y portion must be zero for the number to lie on the X (real number) axis.

Now consider two complex numbers, both with angle zero:

$Z_1 = R_1 \text{ angle } (0), Z_2 = R_2 \text{ angle } (0)$ with the product
$Z_1 \cdot Z_2 = R_1 \cdot R_2 \text{ angle } (0 + 0) = R_1 \cdot R_2 \qquad$ (reverts to traditional multiplication)

Now consider another set of complex numbers, both with angles of 180°:

$$Z_1 = R_1 \text{ angle } (180), Z_2 = R_2 \text{ angle } (180)$$

From the relation above, $Y = R \cdot \sin(\Omega)$, $X = R \cdot \cos(\Omega)$. With angles of 180°, $Y = 0$, $X = -R$, meaning the point Z lies on the negative part of the real axis. Z is simply a negative real number. If we multiply two real negative numbers, we know that we should get a real positive number. Let's check using complex multiplication:

$$Z_1 \cdot Z_2 = R_1 \cdot R_2 \text{ angle } (180 + 180) = R_1 \cdot R_2 \text{ angle } (360).$$

The angle 360° is all the way around the circle and equal to 0°:

$$Z_1 \cdot Z_2 = R_1 \cdot R_2 \text{ angle } (360) = R_1 \cdot R_2 \text{ angle } (0) = R_1 \cdot R_2 = X_1 \cdot X_2$$

The result is a real positive number.

There is another point to all these exercises, which is to explain why we chose something strange like j equals the square root of –1 to designate the vertical axis in the complex number plane. Be patient; we're almost there.

There are four "special" angles: 0°, 90°, 180°, 270°. Notice that

    $Z = R$ angle $(0) = X$ degrees is a positive real number on a positive real axis.
    $Z = R$ angle $(90) = Y$ degrees is a positive imaginary number on a positive
                   imaginary axis.
    $Z = R$ angle $(180) = -X$ degrees is a negative real number on a negative real axis.
    $Z = R$ angle $(270) = -Y$ degrees is a negative imaginary number on a negative
                  imaginary axis.

If we add 360° to any complex number, it wraps all the way around the circle. Or we can have a negative angle, which means just going backward (clockwise) around the circle:

    $Z = R$ angle $(0) = R$ angle $(360) = R$ angle $(720)\dots$
    $Z = R$ angle $(120) = R$ angle $(480) = R$ angle $(840)\dots$
    $Z = R$ angle $(90) = R$ angle $(-270) = R$ angle $(-630)\dots$
    $Z = R$ angle $(-53) = R$ angle $(307) = R$ angle $(667)\dots$

We now know when multiplying two complex numbers, the magnitudes R are multiplied and the angles Ω are summed. Now let's consider a few sample cases to illustrate how this imaginary "j" operator helps us.

Imagine two complex numbers with only an imaginary component. They are both located on the positive imaginary axis:

$$Z_1 = jY_1, Z_2 = jY_2 \quad \text{(real parts } X_1 = X_2 = 0)$$
$$Z_1 \cdot Z_2 = jY_1 \cdot jY_2 = j \cdot j \cdot Y_1 \cdot Y_2$$

Recall, we defined j = sqrt(–1), so $j^2 = -1$

$$Z_1 \cdot Z_2 = -(Y_1 \cdot Y_2), \text{ a negative real number}$$

Or equivalently,

$$Z_1 = R_1 \text{ angle } (90), \ Z_2 = R_2 \text{ angle } (90)$$
$$Z_1 \cdot Z_2 = R_1 \cdot R_1 \text{ angle } (180) = -(R_1 \cdot R_2) = -(Y_1 \cdot Y_2) \text{ since } X_1 = X_2 = 0$$

You can experiment with other combinations, but what you will find is that the arithmetic of adding angles around the circle when multiplying complex numbers works out perfectly when we designate the positive imaginary axis with j and the negative imaginary axis with –j.

By visualizing this business of going around the circle, you can see by inspection that

- Multiply two positive real numbers, both angles = 0, the result has an angle of zero:

$$3 \cdot 5 = 15$$

- Multiply two negative real numbers, both angles = 180 (or –180), the result has an angle of zero (or 360):

$$-3 \cdot -5 = 15$$

- Multiply a positive real number (angle 0) with a negative real number (angle 180), the result has an angle of 180, a negative real number:

$$3 \cdot -5 = -15$$

- Multiply a positive real number (angle 0) with a positive imaginary number (angle 90), the result has an angle of 90, an imaginary number:

$$j3 \cdot 5 = j15$$

- Multiply a positive imaginary number (angle 0) with a positive imaginary number (angle 90), the result has an angle of 180, a negative real number:

$$j3 \cdot j5 = j^2 \cdot 15 = -15$$

- Multiply a negative imaginary number (angle –90) with a negative imaginary number (angle –90), the result has an angle of 180, a negative real number:

$$-j3 \cdot -j5 = (-j) \cdot (-j) \cdot 15 = -(-(-(15))) = -15$$

- Multiply a positive imaginary number (angle 90) with a negative imaginary number (angle –90), the result has an angle of 0, a positive real number:

$$j3 \cdot -j5 = j \cdot (-j) \cdot 15 = -(-(15)) = 15$$

2.3 Complex Conjugate

The last example illustrates a special case. Every number $Z = R$ angle (Ω) has what is called a *complex conjugate*, $Z^* = R$ angle $(-\Omega)$. In the earlier example, $\Omega = 90$, but Ω can be any angle. The * symbol is the complex conjugate symbol, and means to take the point Z and mirror it across the X axis as shown in Figure 2.4.

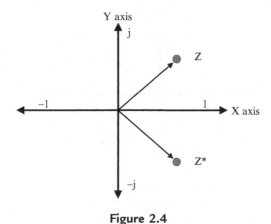

Figure 2.4

So $Z = X + jY$ has the conjugate $Z^* = X - jY$. We just negate or reverse the sign of the imaginary part of a number to get its conjugate, or if in polar form, just negate the sign of the angle:

$$Z = R \text{ angle } (\Omega), Z^* = R \text{ angle } (-\Omega)$$

A special property of the complex conjugate is that for any complex number

$$Z \cdot Z^* = R \text{ angle } (\Omega) \cdot R \text{ angle } (-\Omega) = R^2 \text{angle}(0).$$

In other words, when you multiply a number by its conjugate, the product is a real number, equal to the magnitude squared. This point will become important in digital signal processing because it can be used to compute the power of a complex signal.

To summarize, we have tried to show that the imaginary numbers that are used to form things called *complex* numbers are really not so complex, and *imaginary* is really a very misleading description. What we have really been trying to do is to create a two-dimensional number plane and define a set of expanded arithmetic rules to manipulate the numbers in it. Now we are ready to move onto the next topic, the complex exponential.

2.4 The Complex Exponential

The complex exponential has an intimidating sound to it, but in reality, it is very simple to visualize. It is simply the unit circle (radius $= 1$) on the complex number plane (see Figure 2.5).

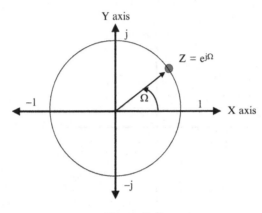

Figure 2.5

Any point on the unit circle can be represented by "e" or raised to the power (j · angle) or also expressed as $e^{j\Omega}$, which is called a complex exponential function. A few examples should help.

Let the angle $\Omega = 0°$. Anything raised to the power 0 is equal to 1. This checks out, since this is the point 1 on the positive real axis.

Let angle $\Omega = 90°$. The complex exponential is e^{j90}. This is the point j on the positive imaginary axis. We need a way to evaluate the complex exponential to show this point. This leads to the Euler equation. This equation can easily be derived using Taylor series expansion for exponentials, but we've promised to minimize the math. The result is

$$e^{j\Omega} = \cos(\Omega) + j\sin(\Omega)$$

Let's try exp(j90) again. Using the Euler equation, we get

$$e^{j90} = \cos(90) + j\sin(90) = 0 + j \cdot 1 = j$$

Imagine the point $Z = e^{j\Omega}$ with the angle Ω starting at 0° and gradually increasing to 360°. This will start at the point $+1$ on the real axis and move counterclockwise around the circle until it ends up where it started, at 1 again. If the angle starts at 0 and gradually decreases until it reaches -360, the point will do exactly the same thing, except rotate in a clockwise fashion.

Now we know from the Euler equation that the complex exponential has a real and an imaginary component. Try to imagine the movement of the point on the unit circle as reflected on the real axis (imagine a second point, allowed to move only on the real axis, trying to follow the first point as it moves about the circle). The movement of the second point on the real axis will equal to $\cos(\Omega)$. So if we continually rotate in either direction about the unit circle, the real component will move back and forth between $+1$ and -1 using the motion of the cosine function. Similarly, the movement of the point on the unit circle as reflected on the imaginary axis will be similar, except instead of starting at a value of $+1$, it will start with a value of 0. The pattern of motion will lag by 90°. The imaginary axis movement is equal to $j \cdot \sin(\Omega)$, and the imaginary component will move back and forth between j and $-j$ using the sine function.

This scenario is shown in Figure 2.6, where the dashed line represents the imaginary axis movement of $j \cdot \sin(\Omega)$ and the dotted line represents the real axis movement of $\cos(\Omega)$.

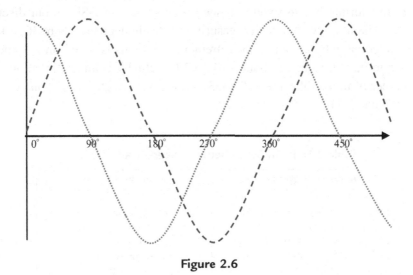

Figure 2.6

$$X = \cos(\Omega) \quad \text{(real axis movement)}$$
$$Y = \sin(\Omega) \quad \text{(imaginary axis movement)}$$

This gives us a better way to express a complex number in polar coordinates:

Recall $Z = X + jY = R$ angle (Ω)

As we saw before,

$$X = \text{Rcos}\,(\Omega)$$
$$Y = \text{Rsin}\,(\Omega)$$

So we can see that angle (Ω) has the same meaning as $\exp(j\Omega)$. Also, for the unit circle, $R = 1$ by definition. So our new way to express a number in polar form using the complex exponential is

$$Z = R\,\text{angle}\,(\Omega) = R\,e^{j\Omega} \quad \text{(any point in a complex plane)}$$
$$Z = \text{angle}\,(\Omega) = e^{j\Omega} \quad \text{(for } R = 1, \text{ any point on the unit circle)}$$

This is the way you will often see these expressions in the textbooks and industry literature.

2.5 Measuring Angles in Radians

The last curve ball in this chapter involves measuring angles in radians. You will have to get accustomed to this format because you will see it everywhere in DSP. In our discussion, to make things more familiar, we will start measuring angles in degrees, where 360° describes a full circle. More commonly, the angle measurement in radians is based on π, which is a number defined to have a value of about 3.141592 (it actually is an irrational number, with an infinite number of digits, like $1/3 = 0.3333...$). It takes exactly 2π radians to describe a full circle (see Table 2.1).

Table 2.1: Equivalence between degrees and radians

Angle in Degrees	Angle in Radians
0	$0\,\pi$
45	$\pi/4$
90	$\pi/2$
$180 = -180$	$\pi = -\pi$
$270 = -90$	$(3/2)\,\pi = -\,\pi/2$
$360 = 0$	$2\pi = 0\pi$
$-360 = 0$	$-2\pi = 0\pi$
$540 = 180$	$3\pi = \pi$
$-540 = -180 = 180$	$-3\pi = -\pi = \pi$

Just as angle measurements are periodic in 360°, they are also periodic in 2π radians. Using π is really no different than getting used to meters rather than using feet for measuring distances (or the reverse if you didn't grow up in the United States).

We are going to see this same concept later in sampling theory, where everything tends to wrap around, or behave periodically. We can visualize this concept as traveling either clockwise (negative rotation) or counterclockwise (positive rotation) around the circle.

There is one more DSP convention to be aware of. The real component (we used X in the preceding discussion) is usually called the "I," or in-phase component, and the imaginary component (we used Y in the preceding discussion) is usually referred to as the "Q," or quadrature phase component. In many DSP algorithms, the digital signal processing must be performed simultaneously on both I and Q data streams, which we now know simply represent the signal's movement, over time, within the two dimensions of our complex number plane.

CHAPTER 3

Sampling, Aliasing, and Quantization

Now that we have the basic background material covered, let's start talking about DSP. The starting point we need to understand is sampling and its effect on the signal of interest. To take an analog signal and convert it to a digital signal, we need to sample the signal using a device called an analog-to-digital converter (ADC). The ADC measures the signal at rapid intervals, and these measurements are called samples. It outputs a digital signal proportional to the amplitude of the analog signal at that instant. This scenario can be compared to looking at an object with only a strobe light for illumination. You can see the object only when the strobe light flashes. If the object is not moving, then everything looks pretty much the same, as if we had used a normal, continuous light source. Things get interesting when we look at a moving object with the strobe light. If the object is moving rapidly, then the appearance of the motion can be quite different from that viewed under normal light. We can also see strange effects even if the object is moving fairly slowly, and we reduce the rate of the strobe light enough. Intuitively, we can see that what is important is the rate of the strobe light compared to the rate of movement of the illuminated object. As long as the light is strobing fast compared to the movement of the object, the movement of the object looks very fluid and normal. When the light is strobing slowly compared to the rate of object movement, the movement of the object looks funny, often like slow motion, because we can see the object is moving, but we miss the sense of continuous, fluid movement.

Let's examine one more example many of us experienced as children. Imagine trying to make your own animated movie and sketching a character on index cards. We want to depict this character moving, perhaps jumping and falling. We might sketch 20 or 40 cards, each showing the same character in sequential stages of this motion, with just small movement changes from one card to the next. When we are finished, we can show it to our friends by holding one edge of the deck of index cards and flipping through it quickly by thumbing the other edge. We see our character in this continuous motion of jumping and falling by flipping through the deck of index cards in a second or two.

Actually, whenever we watch TV, this same thing is occurring. But the TV is updating the screen at about 60 times per second, which is rapid enough that we don't notice the separate frames, and we think that we are seeing continuous motion.

So it makes sense that if we sample a signal very fast compared to how rapidly the signal is changing, we get a pretty accurate sampled representation of the signal, but if we sample too slow, we see a distorted version of the signal.

Digital Signal Processing 101. DOI: 10.1016/B978-1-85617-921-8.00007-9

The graphs in Figures 3.1 and 3.2 show two different sinusoidal signals being sampled. The slower moving signal (lower frequency) in Figure 3.1 can be represented accurately with the indicated sample rate, but the faster moving signal (higher frequency) in Figure 3.2 is not accurately represented by our sample rate. In fact, it actually appears to be a slow-moving (low frequency) signal, as indicated by the dashed line. This shows the importance of sampling fast enough for a given frequency signal.

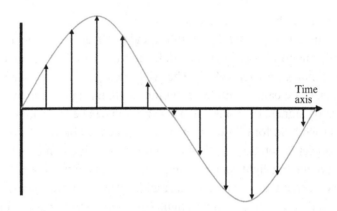

Figure 3.1:
Sampling a low-frequency signal (arrows indicate sample instants).

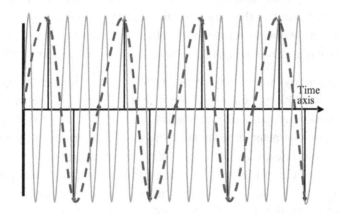

Figure 3.2:
Sampling a high-frequency signal (same sample instants).

The dashed line in Figure 3.2 shows how the sampled signal will appear if we connect the sample dots and smooth out the signal. Notice that since the actual (solid line) signal is changing so rapidly between sampling instants, this movement is not apparent in the sampled version of the signal. The sampled version actually appears to be a lower frequency signal than the actual signal. This effect is known as aliasing.

We need a way to quantify how fast we must sample to accurately represent a given signal. We also need to better understand exactly what is happening when aliasing occurs. The idea may seem strange, but in some instances aliasing can be useful.

Let's go back to the analogy of the strobe light and try another thought experiment. Imagine a spinning wheel with a single dot near the edge. Let's set the strobe light to flash every 1/8 of a second, or 8 times per second. Figures 3.3 through 3.10 show what we see over six flashes, depending on how fast the wheel is rotating. Time increments from left to right in all figures.

Wheel rotating counterclockwise once per second, or 1 Hz

Figure 3.3:
The dot moves 1/8 of a revolution, or $\pi/4$ radians with each strobe flash.

Wheel rotating counterclockwise twice per second, or 2 Hz

Figure 3.4:
Now the dot moves twice as fast, ¼ of a revolution, or $\pi/2$ radians with each strobe flash.

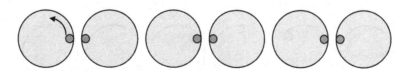

Wheel rotating counterclockwise 4 times per second, or 4 Hz

Figure 3.5:
The dot moves ½ of a revolution, or π radians with each strobe flash. It appears to alternate on each side of the circle with each flash. Can you be sure which direction the wheel is rotating?

Wheel rotating counterclockwise 6 times per second, or 6 Hz

Figure 3.6:
The dot moves counterclockwise 3/4 of a revolution, or 3/2π radians with each strobe flash. But it appears to be moving backward (clockwise).

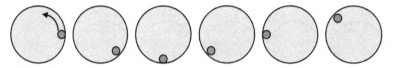

Wheel rotating counterclockwise 7 times per second, or 7 Hz

Figure 3.7:
The dot moves almost a complete revolution counterclockwise, 7/4π radians with each strobe flash. Now it definitely appears to be moving backward (clockwise).

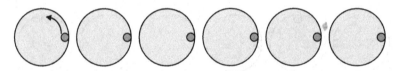

Wheel rotating counterclockwise 8 times per second, or 8 Hz

Figure 3.8:
It looks as though the dot stopped moving! Here, the dot completes exactly one revolution (2π radians) every strobe interval. You can't tell whether the wheel is moving at all, spinning forward or backward. In fact, could the wheel be rotating twice per strobe interval (4π radians)?

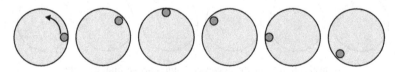

Wheel rotating counterclockwise 9 times per second, or 9 Hz

Figure 3.9:
The image sure looks the same as when the wheel was rotating once per second, or 1 Hz. In fact, the wheel is moving 9/4π radians with each strobe flash.

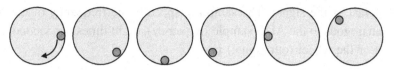

Wheel rotating backward (clockwise) once per second, or −1 Hz

Figure 3.10:
Now we stopped the wheel and started rotating backward at $-\pi/4$ radians with each strobe flash. Notice that this appears exactly the same as when we were rotating forward at $7/4\pi$ radians with each strobe flash.

Doesn't all this look familiar from the preceding chapter? The positive rotation is wrapping around and appears like a negative rotation as the wheel speed increases. And the rotation perception appears periodic in 2π radians per strobe, just like the angle measurements.

The reality is that once we sample a signal (this is what we are doing by flashing the strobe light), we cannot be sure what has happened in between flashes. Our natural instinct is to assume that the signal (or dot in our example) took the shortest path from where it appears in one flash and then in the subsequent flash. But as we can see in the preceding figures, this assumption can be misleading. The dot could be moving around the circle in the opposite direction (taking the longer path) to get to the point where we see it on the next flash. Or imagine that the wheel is rotating in the assumed direction, but it rotates one full revolution plus "a little bit extra" every flash (the 9 Hz diagram). What we see is only the "a little bit extra" on every flash. For that matter, the wheel could go around 10 times plus the same "a little bit extra," and we could not tell the difference.

3.1 Nyquist Sampling Rule

To prevent this confusion, we have to come up with a sampling rule, or convention. What we are going to agree is that we will always sample (or strobe) at least twice as fast as the frequency of the signal we are interested in. And in reality, we need to have some margin, so we had better make sure we are sampling *more* than twice as fast as our signal. Going back to the preceding example, at what point do things start to get fishy?

Consider what happens when we start the wheel moving slowly in a counterclockwise direction. Everything looks fine until we reach a rotational speed of 4 Hz. At this point the dot will appear to be alternating on either side of the circle with each strobe flash. Once we have reached this point, we can no longer tell which direction the wheel is rotating; it will look the same rotating both directions. This is the critical point, where we are sampling at exactly twice as fast as the signal. The sampling speed is the frequency of the strobe light (this would be analogous to the ADC sample frequency), eight times per second, or 8 Hz. The rotational speed of the wheel (our signal) is 4 Hz.

If we spin the wheel any faster, it appears as though it begins to move backward (clockwise), and by the time we reach a rotational speed of 8 Hz, it appears to stop altogether. Spinning still faster will appear as though the wheel moves forward again, until it again appears to start going backward and the cycle repeats.

To summarize, whenever you have a sampled signal, you cannot really be sure of its frequency. But if you assume that the rule was followed—that the signal was sampled at more than twice the frequency of the signal itself—then the sampled signal will really represent the same frequency as the actual signal prior to sampling. The critical frequency that the signal must not ever exceed, which is one half of the sampling frequency, is called the *Nyquist* frequency.

If we follow this rule, then we can avoid the aliasing phenomenon we demonstrated with the moving wheel example earlier. Normally, the ADC converter frequency is selected to be high enough to sample the signal on which we want to perform digital signal processing. To make sure that unwanted signals above the Nyquist frequency do not enter the ADC and cause aliasing, the desired signal in usually passed through an analog low-pass filter, which attenuates any unwanted high-frequency content signals, just prior to the ADC.

A common example is the telephone system. Our voices are assumed to have a maximum frequency of about 3600 Hz. At the microphone, our voice is filtered by an analog filter to eliminate, or at least substantially reduce, any frequencies above 3600 Hz. Then the microphone signal is sampled at 8000 Hz, or 8 kHz. All subsequent digital signal processing occurs on this 8 kSPS (kilo-samples per second) signal. That is why if you hear music in the background while on the telephone, the music will sound flat or distorted. Our ears can detect up to about 15 kHz frequencies, and music generally has frequency content exceeding 3600 Hz. But little of this higher frequency content will be passed through the telephone system.

In the next chapter, we will start representing our signals and sampling in the frequency (or spectral) domain. This means that when we plot the signal spectrum, the X axis will represent increasing frequency (see Figure 3.11).

So far, we have covered the most important effects of sampling, but there remains one last issue related to sampling: quantization. We saw earlier how a digital signal is represented numerically and how we must be sure that the numbers representing the sampled signal do not exceed the range of our binary or hexadecimal number range.

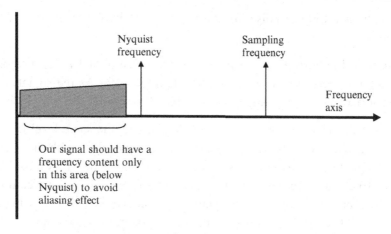

Figure 3.11

3.2 Quantization

What happens if the signals are very small? Remember in our discussion of signed fractional 8-bit fixed-point numbers, the range of values we could represent was from −1 to +1 (well, almost +1). The step size was 1/128, which works out to 0.0078125. Let's say the signal has an actual value of 0.5030 at the instant in time we sample it, using an 8-bit ADC. How closely can the ADC represent this value? Let's compare this to a signal that is 1/10 the level of our first sample, or 0.0503. And again, consider a signal with a value 1/10 the level of the second sample, at 0.00503. The following table shows the possible outputs from an 8-bit ADC at each of these signal levels and the error that will result in the conversion of the actual signal to sampled signal value. We say "possible outputs" because we are assuming that the ADC will output either the value immediately above or below the actual input signal level.

Table 3.1: Quantization effect with 8-bit signal

Signal Level	Closest 8-Bit Representation	Hexadecimal Value	Actual Error	Error as a Percent of Signal Level
0.50300	0.5000000	0x40	0.00300	0.596%
0.50300	0.5078125	0x41	−0.0048128	0.957%
0.05030	0.0468750	0x06	0.003425	6.809%
0.05030	0.0546875	0x07	−0.0043875	8.722%
0.00503	0.000000	0x00	0.00503	100%
0.00503	0.0078125	0x01	−0.0027825	55.32%

The actual error level remains more or less in the same range over the different signal ranges. This error level will fluctuate, depending on the exact signal value, but with our 8-bit signed example, the error level will always be less than 1/128, or 0.0087125. This fluctuating error signal will be seen as a form of noise, or unwanted signal, by the DSP system. It is called quantization noise.

When the signal level is fairly large for the allowable range (0.503 is close to one half the maximum value), the percentage error is small = less than 1%. As the signal level gets smaller, the error percentage gets larger, as the table shows.

The quantization noise is always present and is, on average, the same level (any noise-like signal will rise and fall randomly, so we usually concern ourselves with the average level). But as the input signal decreases in level, the quantization noise becomes more significant in a relative sense. Eventually, for very small input signal levels, the quantization noise can become so significant that it degrades the quality of whatever signal processing is to be performed. Think of it as like static on a car radio. As you get farther from the radio station, the radio signal gets weaker, and eventually the static noise makes it difficult or unpleasant to listen to, even if you increase the volume.

So what can we do if our signal sometimes is strong (e.g., 0.503) and other times is weak (e.g., 0.00503)? Another way of saying this is that the signal has a large dynamic range. The dynamic range describes the ratio between the largest and smallest value of the signal—in this case, 100.

Suppose we exchange our 8-bit ADC for a 16-bit ADC? Then our maximum range is still from −1 to +1, but our step size is now 1/32,768, which works out to 0.000030518. Let's make a 16-bit table similar to the 8-bit example.

Table 3.2: Quantization effects with 16-bit signal

Signal Level	Closest 16-Bit Representation	Hexadecimal Value	Actual Error	Error as a Percent of Signal Level
0.50300	0.5029907	0x4062	0.000009277	0.00185%
0.50300	0.5030212	0x4063	−0.00002124	0.00422%
0.05030	0.0502930	0x0670	0.00000703	0.0140%
0.05030	0.0503235	0x0671	−0.0000235	0.0467%
0.00503	0.005005	0xA4	0.0000251	0.499%
0.00503	0.0050354	0xA5	−0.0000054	0.107%

What a difference! The actual error is always less than our step size, 1/32,768. But the error as a percent of signal level is dramatically improved. This is what we usually care about in signal processing. Because of the much smaller step size of the 16-bit ADC, the quantization noise is much less, allowing even small signals to be represented with very good precision (<1%). Notice that even for our small signal level, 0.00503, the error is about 0.1%.

Another way of describing this scenario is to introduce the concept of signal-to-noise power ratio, or SNR. This ratio describes the power of the largest signal compared to the background noise. This can be very easily seen on a frequency domain or spectral plot of a signal. There can be many sources of noise, but for now, we are considering only the quantization noise introduced by the ADC sampling.

SNR is usually expressed in decibels (denoted dB), using a logarithmic scale. The SNR of an ideal ADC can be determined by the following equation:

$$\text{SNR}_{\text{quantization}}(\text{dB}) = 6.02^*(\text{number of bits}) + 1.76$$

Basically, for each additional bit of the ADC, 6 dB of SNR is gained. An 8-bit ADC is capable of representing a signal with an SNR of about 48 dB, a 12-bit ADC can do better at 72 dB, and a 16-bit ADC will give up to 96 dB. This accounts only for the effect of quantization noise; in practice there are other effects that also will degrade SNR in a system.

There is one last important point regarding decibels. They are very commonly used in many areas of digital signal processing subsystems. A decibel is simply a signal power ratio, similar to a percentage. But because of the extremely high ratios commonly used (a billion is not uncommon), it is convenient to express this amount logarithmically. The logarithmic expression also allow chains of circuits or signal processing operations, each with its own ratio (e.g., of output power to input power), to simply be added up to find the final ratio.

Where people commonly get confused is in differentiating between signal levels or amplitude (voltage if an analog circuit) and signal power. Power measurements are virtual in the digital world but can be directly measured in analog circuits in which DSP systems interface with.

Two definitions of dB are commonly used:

$$\text{dB}_{\text{voltage}} = \text{dB}_{\text{digital value}} = 20 \cdot \log(\text{voltage signal 1/voltage signal 2})$$

$$\text{dB}_{\text{power}} = 10 \cdot \log(\text{power signal 1/power signal 2})$$

The designations "signal 1" and "signal 2" depend on the situation. For an RF power amplifier, the dB of gain will be 10 log (output power/input power). For an ADC, the dB of SNR will be 20 log (maximum input signal/quantization noise signal level). For a DAC, the dB of spurious free dynamic range will be 20 log (maximum output signal level/largest unwanted frequency component level generated by DAC circuits).

So dB can refer to many different ratios. But it is easy to get confused whether to use to the multiplicative factor of 10 or 20 without understanding the reasoning behind this.

Voltage squared is proportional to power. If a given voltage is doubled in a circuit, it requires four times as much power. This can be easily derived from the basic Ohm's law equation:

$$\text{Power} = \text{Voltage}^2/\text{Resistance}$$

In many analog circuits, signal power is used because that is what the lab instruments work with, and while different systems may use different resistance levels (which affects voltage), power is universal (however, 50 ohm is the most common standard in most analog systems).

The important point is that since voltage is squared, this effect needs to be taken into account in the computation of logarithmic decibel relation. Remember, $\log x^y = y \log x$. Hence, the multiplication factor of 2 is required for voltage ratios, changing the 10 to a 20.

In the digital world, the concepts of resistance and power do not exist. A given signal has specific amplitude, expressed in a digital numerical system (such as signed fractional or integer, for example).

Understanding dB increases using the two measurement methods is important. Let's look at doubling of the amplitude ratio and doubling of the power ratio:

$$6.02 \text{ dB}_{\text{voltage}} = 6.02 \text{ dB}_{\text{digital value}} = 20 \cdot \log(2/1)$$

$$3.01 \text{ dB}_{\text{power}} = 10 \cdot \log(2/1)$$

This is why shifting a digital signal left 1 bit (multiply by 2) causes a 6 dB signal power increase, and why so often the term *6 dB/bit* is used in conjunction with ADCs, DACs, or digital systems in general.

By the same reasoning, doubling in power to a radio frequency (RF) engineer means a 3 dB increase. This will also impact the entire system. Coding gain, as used with error correcting code methods, is based on power. All signals at antenna interfaces are defined in terms of power, and the decibels used will be power ratios.

In both systems, ratios of equal power or voltage are 0 dB. For example, a unity gain amplifier has a gain of 0 dB:

$$0 \text{ dB}_{\text{power}} = 10 \cdot \log(1/1)$$

A loss would be expressed as a negative dB—for example, a circuit whose output is equal to ½ the input power:

$$-3.01 \text{ dB}_{\text{power}} = 10 \cdot \log(1/2)$$

CHAPTER 4

Frequency Response

When we have a sampled digital signal, we are ready to perform digital signal processing on this signal. In the preceding chapter, we briefly touched on representing the signal in the frequency domain. Usually, our goal is to modify the signal's frequency representation. This is normally performed using a filtering function. This is probably the most fundamental of DSP functions. In the previous example of the telephone system, we talked about sampling the voice signal at 8 kHz. Let's say we want to build an automated system to detect touch tones (on a touch-tone phone, whenever you press a number key, the phone creates two specific tones or frequencies in the audio signal band; i.e., what you hear when pressing the button). We could build a digital filter for each of the possible tones, feed the sampled audio signal into each filter, and monitor the outputs of the filters. In this way, we could detect when the telephone user presses the buttons, and which buttons are pressed, which is exactly what touch-tone phone systems do.

4.1 Frequency Response and the Complex Exponential

In this chapter, we discuss the concept of frequency response. Then in the next chapter, we will develop a way to relate a filter's frequency and time response. First, let's begin with an intuitive way to understand the frequency response of a filter. From the preceding chapter, we learned that we can create a complex exponential signal, which is just a positive or negative frequency rotation about the unit circle (radius = 1). Furthermore, when the frequency of the complex exponential reaches the Nyquist frequency, we have reached the maximum frequency that can be represented for a given sampling rate.

A complex exponential signal has the following form:

$$e^{j\omega t} = \cos(\omega t) + j\sin(\omega t)$$

or

$$e^{j2f\pi} = \cos(2\pi ft) + j\sin(2\pi ft)$$

This is very similar to what we saw before. The earlier angle Ω has been replaced by ωt or by $2\pi ft$. The significance of this is that we are no longer representing a point or vector on the unit circle of the complex number plane (as determined by angle Ω). Now we are

Digital Signal Processing 101. DOI: 10.1016/B978-1-85617-921-8.00008-0

representing a time-varying signal. This signal will move about the unit circle with a rotational speed of ω radians per second. For a given ω (rotational speed), we can determine the position of the signal at any given point in time (denoted by "t").

The second equation is equivalent to the first, except the rotational speed is expressed in cycles (revolutions) per second, denoted by "f." Make sure you understand this concept before moving on because both forms will appear interchangeably in the DSP world:

t   represents time, in seconds
ω   represents rotational speed, in rad/s (2π radians = 1 revolution)
f   represents rotational speed, in cycles (or revolutions) per second

The variables ω and f simply describe the same thing, using different units, like inches and centimeters. Recall that it takes exactly 2π radians to complete a full circle. So 1 cycle/s equals 2π rad/s, or ω = 2πf.

Both ω and f denote the rotational speed in the counterclockwise direction. A negative value for ω or f means we are rotating in the opposite direction (clockwise). Let's now consider an example. Let the rotational speed equal to 1/8 of a circle per second (it takes 8 s to complete 1 revolution). Therefore, f = 1/8. Also, ω = 2πf or ω = $\pi/4$. This signal (let's call it "s") can be described as follows:

$$s(t) = e^{j\pi t/4} = \cos(\pi t/4) + j \sin(\pi t/4)$$

We can evaluate s(t) at any given time t. For example, see Table 4.1.

Table 4.1: Angular motion of complex exponential

Time (s)	$s(t) = e^{j\pi t/4}$	Angle of s(t) (°)	Angle of s(t) (rad)
0	1	0	0
1	0.707 +0.707j	45	$\pi/4$
2	j	90	$\pi/2$
3	−0.707 +0.707j	135	$3\pi/4$
4	−1	180	π
5	−0.707 −0.707j	225	$5\pi/4 = -3\pi/4$
6	−j	270	$3\pi/2 = -\pi/2$
7	0.707 −0.707j	315	$7\pi/8 = -\pi/4$
8	1	360 = 0	$2\pi = 0$

We can also plot s(t). Figure 4.1 shows separate plots of the real (dotted line) part of s(t) and the imaginary (dashed line) part of s(t). On a separate plot in Figure 4.2 is the complete signal s(t) on the unit circle of the complex number plane with time labels at each point.

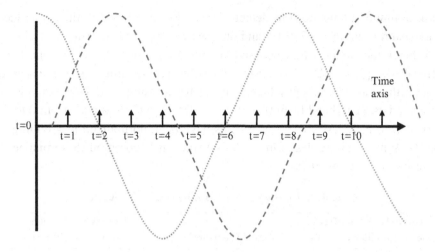

Figure 4.1

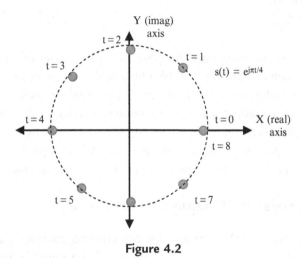

Figure 4.2

4.2 Normalizing Frequency Response

Notice that we are measuring the value of our signal once per second. It turns out that setting the sampling frequency F_s equal to 1 s is often convenient. This process is called normalization. The frequency response of a filter can be expressed as a normalized response, where the input frequency is expressed as a fraction of sampling frequency F_s.

Let's illustrate this concept using the telephone system as an example. In this case, $F_s = 8000$ Hz, $F_{Nyquist} = 4000$ Hz, and one of the touch tones we need to detect is at 770 Hz. We could build a filter with a passband (portion of frequency band to "pass through" the signal, with minimum attenuation) to detect 770 Hz. Since we need a little tolerance, let's make the passband from 760 to 780 Hz and our desired stopbands (portion of frequency band to "stop" or block the signal, with maximum attenuation) from 0 to 760 Hz and from 780 to 4000 Hz (recall $F_{Nyquist} = F_s/2$). Do not forget that we are assuming no signal at frequencies above $F_{Nyquist}$; this has been already taken care of before sampling. This example could also be expressed as a passband from 0.0950 to 0.0975 F_s, and stopbands from 0 to 0.0950 F_s and from 0.0975 to F_s /2. All these frequencies have been normalized by dividing with our $F_s = 8000$ Hz. When you are designing a digital filter, it is common to normalize the frequency response in terms of F_s.

Table 4.2: Mapping filter to normalized frequency

Touch-Tone BandPass Detection Filter Example	Actual Frequency (Hz)	Normalized Frequency ($F_s = 8000$ Hz)
Touch-tone frequency	770	0.09625 F_s (770/8000)
Start of passband	760	0.0950 F_s (760/8000)
End of passband	780	0.0975 F_s (780/8000)
Nyquist frequency	4000	0.50 F_s (4000/8000)

Now let's get back to frequency response. Suppose we input a series of complex exponential signals into our filter. Each exponential will be a little higher frequency than the previous one. We check the output of each frequency. If we input a complex exponential of a given frequency, we will see at the output a complex exponential signal of the same frequency. The reason is that a digital filter is a linear device, so it cannot create new frequencies or change the frequency of a signal passing through it. What it can do is change the amplitude and phase of the input signal. Let's leave the phase part out of it for now and focus on the amplitude.

4.3 Sweeping across the Frequency Response

Suppose we build a digital signal generator that can create a complex exponential signal of any frequency we desire, from 0 to $F_{Nyquist}$. (This is called a numerically controlled oscillator, or NCO, and is common in digital communication systems.) We drive our filter with the NCO, incrementing the frequency in small steps, and we measure the amplitude of the filter output signal. If we plot these data, we will have the magnitude response of our filter with all frequencies from 0 to $F_{Nyquist}$.

If we follow these steps for the touch-tone filter described previously, we will get zero output until we input a complex exponential at frequency 0.0950 F_s. From frequency 0.0950 F_s until 0.0975 F_s, the complex exponential will pass through our filter and could be detected. Above 0.0975 F_s up until $F_{Nyquist}$, we will get zero output. This type of filter is called a bandpass filter because it allows only a portion of the frequency band to pass, and it blocks frequencies above and below this band.

Now remember that we can also have negative frequency complex exponentials, so we could also do a similar plot from 0 to $-F_{\text{Nyquist}}$. For the vast majority of digital filters, the frequency response is the same whether the input is a positive or negative frequency. All filters with real coefficients have this property.

Referencing everything to F_s is also convenient when using our NCO. The NCO does not know that it is being clocked at 8 kHz sampling frequency. Instead, when we program it, we need to set the desired frequency output in terms of F_s. For example, to test our filter with an input of 770 Hz, we need to set the NCO to produce a complex exponential at $0.09625\ F_s$ or $(770/8000)\ F_s$.

4.4 Example Frequency Responses

After this lengthy explanation, let's look at some examples of frequency responses to help clarify things. In this chapter, we depict the frequency response of ideal filters. First, we show low pass, then high pass, and lastly a bandpass, like our touch-tone 770 Hz detector.

Note that the filter response repeats at intervals of F_s. This is an artifact of sampling, as explained in the preceding chapter. The valid portion of the sampled frequency response is from $-F_{\text{Nyquist}}$ to $+F_{\text{Nyquist}}$. Just as we saw the apparent rate of rotation of the wheel with the red dot reach a maximum at a rate of ½ the sample rate and then slowly decrease until it finally stopped when the rate of rotation equaled the sample rate, the frequency response of the filter will behave similarly between zero and F_s. So the filter response near zero will be the same as that near F_s, and the filter response just below F_{Nyquist} will be the same as just above F_{Nyquist}. In fact, since this phenomenon is well understood, there is really no reason to plot a digital filter's response above F_{Nyquist} and below $-F_{\text{Nyquist}}$. We plot it in Figure 4.3 mainly to show this point.

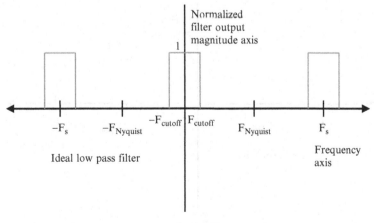

Figure 4.3

The low-pass filter passes frequencies near zero. Both positive and negative frequencies are passed. When the frequency reaches the chosen F_{cutoff}, the filter no longer passes the signal. As we approach any multiple of F_s, we see the aliasing of the filter response, as shown in Figure 4.4.

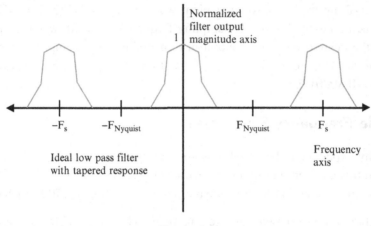

Figure 4.4

In Figure 4.4 we gradually reduced the low-pass filter response as it approaches F_{cutoff}. This example shows that it is possible to design for an arbitrary passband response and a gradual transition to the stopband, not just a flat response in the filter passband with an abrupt transition at F_{cutoff}.

The high-pass filter in Figure 4.5 passes frequencies near $F_{Nyquist}$. Both positive and negative frequencies are passed. When the frequency falls below F_{cutoff}, the filter no longer passes the signal. Aliasing is again seen by the symmetry about $F_{Nyquist}$.

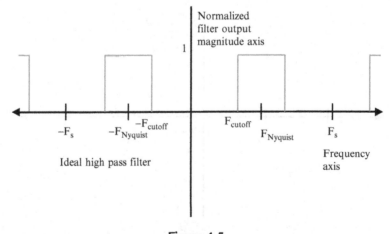

Figure 4.5

Bandpass (do not confuse this with the early term "passband") filters pass only a specific band, or portion of frequencies, that does not include either DC (0 Hz) or $F_{Nyquist}$. The bandpass filter in Figure 4.6 again shows the alias effect. This repeats to infinity at every multiple of F_s. Going back to Figure 4.3, the low-pass filter frequency plot, we see the alias centered at F_s. So according to our plot, the filter will pass frequencies around F_s, even though this is above our F_{cutoff}. So if we input a signal near F_s to an ADC, it will appear as a frequency near zero and be passed by our filter.

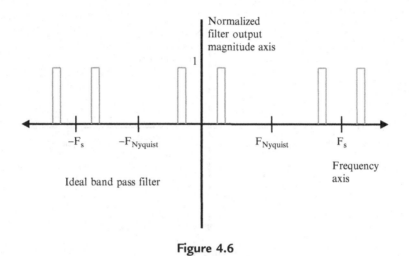

Figure 4.6

This concept is difficult to understand, which is why there is so much repetition on this theme. Do not be worried if you have to go back and forth a few times to review the diagrams or earlier chapters to really satisfy yourself that you understand it. It is a key concept for much of what follows, and unfortunately, some experienced people working in the DSP industry still have trouble with these fundamentals.

4.5 Linear Phase Response

Earlier, we decided to ignore the phase response. The reason is that, for most filters, the phase response is not something we need to worry about. The most common type of digital filter is called the finite impulse response, or FIR, and it has what is called a linear phase response. This means that every frequency passing through the filter experiences the same delay, which works out to a linearly increasing phase as the frequency increases. This sounds complicated, but the short explanation is that for FIR filters, we do not need to worry about phase response. This is the type of filter we will discuss in detail in the next chapter, and is the most common DSP function implemented.

4.6 Normalized Frequency Response Plots

Now we have one more topic to finish out this chapter. We now know that it is sufficient to describe a digital filter response from $-F_{Nyquist}$ to $+F_{Nyquist}$. But frequently, in textbooks or filter design programs, the frequency response is given in terms of normalized radians per second, ω, rather than normalized frequency, as shown in Figure 4.7.

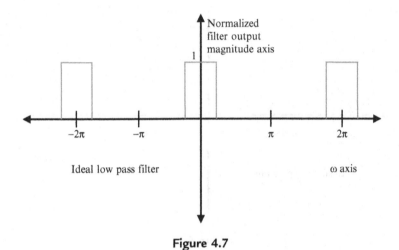

Figure 4.7

Again, let's refer back to our complex exponential, $e^{j\omega t}$. We are sampling at a normalized frequency of 1 sample/s, or 2π rad/s. This means that a complex exponential of frequency π rad/s will correspond to $F_{Nyquist}$, and 2π rad/s will correspond to F_s. This is exactly what we see when we plot the frequency response using the ω axis.

As a review, see the following table, which gives sample low-pass response cutoff both in terms of ω and F_s. It is important to be comfortable going between these two representations because most filter design programs are defined in terms of ω, but to design your filter, you need to be able to translate to the real frequencies used in your application to the filter response in terms of ω.

The cutoff point is defined as a percentage of the maximum bandwidth possible for a low-pass filter, with 100% becoming an all-pass filter (every frequency passes through). Normally, the filter frequency response is symmetric, so $F_{cutoff} = -F_{cutoff}$.

Table 4.3: Converting radians and Hertz

% of Maximum Possible Low-Pass Filter Bandwidth	Ω (rad/s)	F_s (Hz or cycles/s)
10%	$\pi/10$	$F_s/20$
10%	$-\pi/10$	$-F_s/20$
25%	$\pi/4$	$F_s/8$
50%	$\pi/2$	$F_s/4$
80%	$4\pi/5$	$2F_s/5$
100%	π	$F_s/2$ or $F_{Nyquist}$

CHAPTER 5

Finite Impulse Response (FIR) Filters

This chapter focuses on the workhorse of DSP: the finite infinite response (FIR) filter. We discuss three main topics. First, we talk about the structure of an FIR filter, how to build one, and some of its properties. Next, we show how, given the filter coefficients, to compute the frequency response of the filter (normally, this is done with software, but we can gain insight by understanding how to perform this procedure). Last, we show a method to compute coefficients to meet a given frequency response. This last step is what is commonly required of a DSP designer: to find the number and value of filter coefficients required for frequency response required by the application. Again, this procedure is normally performed with software, but we need to gain insight as to what is involved. Filter design is a process in which we cannot have everything, and compromise is necessary. To use software tools to design filters, we need to understand what the different trade-offs are and how they interact.

5.1 FIR Filter Construction

Let's begin with how to construct an FIR filter. An FIR filter is built of multipliers and adders. It can be implemented in hardware or software, and run in a serial fashion, parallel fashion, or some combination. We focus on the parallel implementation because it is the most straightforward to understand.

FIR filters, and DSP in general, often use delay elements. A delay element is simply a clocked register, and a series of delay elements simply comprise a shift register. However, in DSP, a one-sample delay element is often represented as a box with a z^{-1} symbol. This comes from the mathematical properties of the z-transform, which we have not covered in the interest of minimizing mathematics (although there is a discussion of z-transform in Appendix D). But we cannot skip the use of z^{-1} representing a single clock or register delay because it is prevalent in DSP diagrams and literature. We use it here so that you can get accustomed to seeing this representation.

A key property of an FIR filter is the number of taps, or multipliers, required to compute each output. In a parallel implementation, the number of taps equals the number of multipliers. In a serial implementation, one multiplier is used to perform all the multiplication operations sequentially for each output. Assuming single clock cycle multipliers, a parallel FIR filter can produce one output each clock cycle, and a serial FIR filter would require N clock cycles

Digital Signal Processing 101. DOI: 10.1016/B978-1-85617-921-8.00009-2

to produce each output, where N is the number of filter taps. Filters can sometimes have hundreds of taps. Figure 5.1 shows a small five-tap parallel filter.

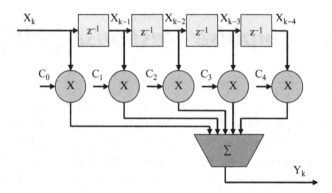

Figure 5.1

The inputs and outputs of the FIR filter are sampled data. For simplicity, let's assume that the inputs, outputs, and filter coefficients C_m are all real numbers. The input data stream is denoted as x_k, and the output is y_k. The "k" subscript is used to identify the sequence of data. For example, x_{k+1} follows x_k, and x_{k-1} precedes x_k. Often for the purpose of defining a steady state response, we assume that the data streams are infinitely long in time, or that k extends from $-\infty$ to $+\infty$.

The coefficients are usually static (meaning they do not change over time) and determine the filter's frequency response.

In equation form, the filter could be represented as

$$y_k = C_0 \cdot x_k + C_1 \cdot x_{k-1} + C_2 \cdot x_{k-2} + C_3 \cdot x_{k-3} + C_4 \cdot x_{k-4}$$

It is just the sum of multipliers. Writing this equation could get pretty tedious as the number of taps gets larger, so the following shorthand summation is often used:

$$y_k = \sum_{i=0 \text{ to } 4} C_i \, x_{k-i}$$

We can also make the equation for any length of filter. To make our filter of length N, we simply replace the 4 (5 − 1 taps) with N − 1:

$$y_k = \sum_{i=0 \text{ to } N-1} C_i \, x_{k-i}$$

Another way to look at this is that the data stream $\ldots x_{k+2}, x_{k+1}, x_k, x_{k-1}, x_{k-2} \ldots$ is sliding past a fixed array of coefficients. At each clock cycle, the data and coefficients are cross-multiplied, and the outputs of all multipliers for that clock cycle are summed to form a single output (this process is also known as dot product).

Then on the next clock cycle, the data is shifted one place relative to the coefficients (which are fixed), and the process repeated. This process is known as convolution.

The FIR structure is very simple, yet it has the capability to create almost any frequency response, given a sufficient number of taps. This structure is very powerful, but unfortunately not at all intuitive. It's somewhat analogous to the brain—a very simple structure of interconnected neurons, yet the combination can produce amazing results. During the rest of the chapter, we try to gain some understanding of how this happens.

Following is an example using actual numbers, to illustrate this process called convolution.

To start, let's define a filter of 5 coefficients $\{C_0,C_1,C_2,C_3,C_4\} = \{1,3,5,3,1\}$. Our x_k sequence is defined as $\{x_0,x_1,x_2,x_3,x_4,x_5\} = \{-1,1,2,1,4,-1\}$ and $x_k = 0$ for $k < 0$ and for $k > 6$ (everywhere else). Let's start by computing $y_{-1} = \sum_{i=0 \text{ to } N-1} C_i\, x_{-1-i}$.

We can see that the subscript on x will be negative for all $i = 0$ to 4. In this example, $y_k = 0$ for $k < 0$. This means that until there is a non-zero input x_k, the output y_k will also be zero. Things start to happen at $k = 0$ because x_0 is the first nonzero input:

$$y_{-1} = \sum_{i=0 \text{ to } N-1} C_i\, x_{0-i} = (1)(0) + (3)(0) + (5)(0) + (3)(0) + (1)(0) = 0$$

$$y_0 = \sum_{i=0 \text{ to } N-1} C_i\, x_{0-i} = (1)(-1) + (3)(0) + (5)(0) + (3)(0) + (1)(0) = -1$$

$$y_1 = \sum_{i=0 \text{ to } N-1} C_i\, x_{0-i} = (1)(1) + (3)(-1) + (5)(0) + (3)(0) + (1)(0) = -2$$

$$y_2 = \sum_{i=0 \text{ to } N-1} C_i\, x_{0-i} = (1)(2) + (3)(1) + (5)(-1) + (3)(0) + (1)(0) = 0$$

$$y_3 = \sum_{i=0 \text{ to } N-1} C_i\, x_{0-i} = (1)(1) + (3)(2) + (5)(1) + (3)(-1) + (1)(0) = 9$$

$$y_4 = \sum_{i=0 \text{ to } N-1} C_i\, x_{0-i} = (1)(4) + (3)(1) + (5)(2) + (3)(1) + (1)(-1) = 19$$

$$y_5 = \sum_{i=0 \text{ to } N-1} C_i\, x_{0-i} = (1)(-1) + (3)(4) + (5)(1) + (3)(2) + (1)(1) = 23$$

$$y_6 = \sum_{i=0 \text{ to } N-1} C_i\, x_{0-i} = (1)(0) + (3)(-1) + (5)(4) + (3)(1) + (1)(2) = 22$$

$$y_7 = \sum_{i=0 \text{ to } N-1} C_i\, x_{0-i} = (1)(0) + (3)(0) + (5)(-1) + (3)(4) + (1)(1) = 8$$

$$y_8 = \sum_{i=0 \text{ to } N-1} C_i\, x_{0-i} = (1)(0) + (3)(0) + (5)(0) + (3)(-1) + (1)(4) = 1$$

$$y_9 = \sum_{i=0 \text{ to } N-1} C_i\, x_{0-i} = (1)(0) + (3)(0) + (5)(0) + (3)(0) + (1)(-1) = -1$$

$$y_{10} = \sum_{i=0 \text{ to } N-1} C_i\, x_{0-i} = (1)(0) + (3)(0) + (5)(0) + (3)(0) + (1)(0) = 0$$

$$y_{11} = \sum_{i=0 \text{ to } N-1} C_i \, x_{0-i} = (1)(0) + (3)(0) + (5)(0) + (3)(0) + (1)(0) = 0$$

$$y_{12} = \sum_{i=0 \text{ to } N-1} C_i \, x_{0-i} = (1)(0) + (3)(0) + (5)(0) + (3)(0) + (1)(0) = 0$$

This procedure is definitely tedious. There are a couple of points to notice. Follow the input $x_4 = 4$ (bold) in our example. See how it moves across, from one multiplier to the next. Each input sample x_k is multiplied by each tap in turn. Once it passes through the filter, that input sample is discarded and has no further influence on the output. In our example, x_4 is discarded after computing y_8.

Once the last nonzero input data x_k has shifted its way through the filter taps, the output data y_k will go to zero (this starts at $k = 10$ in our example).

Now let's consider a special case in which $x_k = 1$ for $k = 0$, and $x_k = 0$ for $k \neq 0$. This means that we have only one nonzero input sample, and it is equal to 1. Now if we again compute the output, which is simpler this time, we get

$$y_{-1} = \sum_{i=0 \text{ to } N-1} C_i \, x_{0-i} = (1)(0) + (3)(0) + (5)(0) + (3)(0) + (1)(0) = 0$$

$$y_0 = \sum_{i=0 \text{ to } N-1} C_i \, x_{0-i} = (1)(1) + (3)(0) + (5)(0) + (3)(0) + (1)(0) = 1$$

$$y_1 = \sum_{i=0 \text{ to } N-1} C_i \, x_{0-i} = (1)(0) + (3)(1) + (5)(0) + (3)(0) + (1)(0) = 3$$

$$y_2 = \sum_{i=0 \text{ to } N-1} C_i \, x_{0-i} = (1)(0) + (3)(0) + (5)(1) + (3)(0) + (1)(0) = 5$$

$$y_3 = \sum_{i=0 \text{ to } N-1} C_i \, x_{0-i} = (1)(0) + (3)(0) + (5)(0) + (3)(1) + (1)(0) = 3$$

$$y_4 = \sum_{i=0 \text{ to } N-1} C_i \, x_{0-i} = (1)(0) + (3)(0) + (5)(0) + (3)(0) + (1)(1) = 1$$

$$y_5 = \sum_{i=0 \text{ to } N-1} C_i \, x_{0-i} = (1)(0) + (3)(0) + (5)(0) + (3)(0) + (1)(0) = 0$$

$$y_6 = \sum_{i=0 \text{ to } N-1} C_i \, x_{0-i} = (1)(0) + (3)(0) + (5)(0) + (3)(0) + (1)(0) = 0$$

Notice that the output is the same sequence as the coefficients. This should come as no surprise when you think about it. This output is defined as the filter's impulse response, named as it occurs when the filter input is an impulse, or a single nonzero input equal to 1. This gives the FIR filter its name. By "finite impulse response," or FIR, this indicates that if this type of filter is driven with an impulse, we will see a response (the output) has a finite length, after which it becomes zero. This point may seem trivial, but it is a very good property to have, as we will see in Chapter 8 on infinite impulse response filters.

5.2 Computing Frequency Response

What we have covered so far is the mechanics of building the filter and how to compute the output data, given the coefficients and input data. But we do not have any intuitive feeling as to how this operation can allow some frequencies to pass through and yet block other frequencies. A very basic understanding of a low-pass filter can be gained by examining the concept of averaging. We all know that if we average multiplication results, we get a smoother, more consistent output, and rapid fluctuations are damped out. A moving average filter is simply a filter with all the coefficients set to 1. The more filter taps, the longer the averaging, and the more smoothing takes place. This gives an idea of how a filter structure can remove high frequencies, or rapid fluctuations. Now imagine if the filter taps were alternating +1, −1, +1, −1,... and so on. A slowly varying input signal has adjacent samples nearly the same, and these cancel in the filter, resulting in a nearly zero output. This filter blocks low frequencies. On the other hand, an input signal near the Nyquist rate has big changes from sample to sample, and results in a much larger output. However, to get a more precise handle on how to configure the coefficient values to get the desired frequency response, we need to use a bit of math.

We start by computing the frequency response of the filter from the coefficients. Remember, the frequency response of the filter is determined by the coefficients (also called the impulse response).

Let us begin by trying to determine the frequency response of a filter by measurement. Imagine if we take a complex exponential signal of a given frequency and use this as the input to our filter. Then we measure the output. If the frequency of the exponential signal is in the passband of the filter, it will appear at the output. But if the frequency of the exponential signal is in the stopband of the filter, it will appear at the output with a much lower level than the input, or not at all. Imagine we start with a very low frequency exponential input and do this measurement; then we slightly increase the frequency of the exponential input, measure again, and keep going until the exponential frequency is equal to the Nyquist frequency. If we plot the level of the output signal across the frequency from 0 to $F_{Nyquist}$, we will have the frequency response of the filter. It turns out that we do not have to do all these measurements; instead, we can compute this fairly easily. The computation is as follows:

$y_k = \sum_{i=0 \text{ to } 4} C_i\, x_{k-i}$ This is the output of our five-tap sample filter.

$y_k = \sum_{i=-\infty \text{ to } \infty} C_i\, x_{k-i}$ Same equation, except that we are allowing an infinite number of coefficients (no limits on filter length).

$x_m = e^{j\omega m} = \cos(\omega m) + j\sin(\omega m)$ This is our complex exponential input at ω radians per sample.

Let's take a closer look at the last equation and review a bit.

This last equation is just a sampled version of a signal rotating around the unit circle. We sample at time $= m$ and then sample again at time $= m + 1$. So from one sample to the next, our sampled signal will move ω radians around the unit circle. If we are sampling at 10 times faster than we are moving around the unit circle, then it will take 10 samples to get around the circle, and move $2\pi/10$ radians each sample:

$$x_m = e^{j2\pi m/10} = \cos(2\pi m/10) + j\sin(2\pi m/10), \quad \text{when } \omega = 2\pi/10$$

To clarify, the following table shows $x_m = e^{j2\pi m/10}$ evaluated at various m. If you want to check using a calculator, remember that the angles are in units of radians, not degrees.

We could next increase x_m so that we rotate the unit circle every five samples. This is twice as fast as before. I hope you are getting more comfortable with complex exponentials.

$$x_m = e^{j2\pi m/5} = \cos(2\pi m/5) + j\sin(2\pi m/5), \quad \text{when } \omega = 2\pi/5$$

Table 5.1

$m = 0$	$x_0 = e^{j0} = \cos(0) + j\sin(0)$	$1 + j0$
$m = 1$	$x_1 = e^{j\pi/5} = \cos(\pi/5) + j\sin(\pi/5)$	$0.8090 + j0.5878$
$m = 2$	$x_2 = e^{j2\pi/5} = \cos(2\pi/5) + j\sin(2\pi/5)$	$0.3090 + j0.9511$
$m = 3$	$x_3 = e^{j3\pi/5} = \cos(3\pi/5) + j\sin(3\pi/5)$	$-0.3090 + j0.9511$
$m = 4$	$x_4 = e^{j4\pi/5} = \cos(4\pi/5) + j\sin(4\pi/5)$	$-0.8090 + j0.5878$
$m = 5$	$x_5 = e^{j\pi} = \cos(\pi) + j\sin(\pi)$	$-1 + j0$
$m = 6$	$x_6 = e^{j6\pi/5} = \cos(6\pi/5) + j\sin(6\pi/5)$	$-0.8090 - j0.5878$
$m = 7$	$x_7 = e^{j7\pi/5} = \cos(7\pi/5) + j\sin(7\pi/5)$	$-0.3090 - j0.9511$
$m = 8$	$x_8 = e^{j8\pi/5} = \cos(8\pi/5) + j\sin(8\pi/5)$	$0.3090 - j0.9511$
$m = 9$	$x_9 = e^{j9\pi/5} = \cos(9\pi/5) + j\sin(9\pi/5)$	$0.8090 - j0.5878$
$m = 10$	$x_{10} = x_0 = e^{j2\pi} = \cos(2\pi) + j\sin(2\pi)$	$1 + j0$

Now we go back to the filter equation and substitute the complex exponential input for x_{k-i}.

$y_k = \sum_{i=-\infty \text{ to } \infty} C_i x_{k-I}$

$x_m = e^{j\omega m} = \cos(\omega m) + j\sin(\omega m)$ .... insert k−i for m

$x_{k-i} = e^{j\omega(k-i)} = \cos(\omega(k-i)) + j\sin(\omega(k-i))$ .... next, replace in x_{k-i} in filter equation

$y_k = \sum_{i=-\infty \text{ to } \infty} C_i e^{j\omega(k-i)}$

There is a property of exponentials that we frequently need to use:

$$e^{(a+b)} = e^a \cdot e^b \text{ and } e^{(a-b)} = e^a \cdot e^{-b}$$

If you remember your scientific notation, this makes sense. For example,

$$10^2 \cdot 10^3 = 100 \cdot 1000 = 100,000 = 10^5 = 10^{(2+3)}$$

Now we go back to the filter equation:

$$y_k = \sum_{i=-\infty \text{ to } \infty} C_i\, e^{j\omega(k-i)} = \sum_{i=-\infty \text{ to } \infty} C_i\, e^{j\omega k} \cdot e^{-j\omega i}$$

Let's do a little algebra trick. Notice the term $e^{j\omega k}$ does not contain the term i used in the summation. So we can pull this term out in front of the summation:

$$y_k = e^{j\omega k} \cdot \sum_{i=-\infty \text{ to } \infty} C_i\, e^{-j\omega i}$$

Notice the term $e^{j\omega k}$ is just the complex exponential we used as an input:

$$y_k = x_k \cdot \sum_{i=-\infty \text{ to } \infty} C_i\, e^{-j\omega i}$$

Voila! The expression $\sum_{i=-\infty \text{ to } \infty} C_i\, e^{-j\omega i}$ gives us the value of the frequency response of the filter at frequency ω. It is solely a function of ω and the filter coefficients.

This expression applies a gain factor to the input, x_k, to produce the filter output. Where this expression is large, we are in the passband of the filter. If this expression is close to zero, we are in the stopband of the filter.

Let us give this expression a less cumbersome representation. Again, it is a function of ω, which we expect because the characteristics of the filter vary with frequency. It is also a function of the coefficients, C_i, but these are assumed fixed for a given filter:

$$\text{Frequency response} = H(\omega) = \sum_{i=-\infty \text{ to } \infty} C_i e^{-j\omega i}$$

Now in reality, this equation is not as bad as it looks. The preceding is the generic version of the equation, where we must allow for an infinite number of coefficients (or taps). But suppose we are determining the frequency response of our five-tap sample filter:

$$H(\omega) = \sum_{i-0 \text{ to } 4} C_i\, e^{-j\omega i} \text{ and } \{C_0, C_1, C_2, C_3, C_4\} = \{1, 3, 5, 3, 1\}$$

Let's find the response of the filter at a couple of different frequencies. First, let $\omega = 0$. This corresponds to DC input; we are putting a constant level signal into the filter. This would be $x_k = 1$ for all values k:

$$H(0) = C_0 + C_1 + C_2 + C_3 + C_4 = 1 + 3 + 5 + 3 + 1 = 13$$

This one was simple, since $e^0 = 1$. The DC or zero frequency response of the filter is called the gain of the filter. Often, it may be convenient to force the gain $= 1$, which would involve dividing all the individual filter coefficients by $H(0)$. The passband and stopband characteristics are not altered by this process, since all the coefficients are scaled equally. It just normalizes the frequency response so the passband has a gain equal to 1.

Now compute the frequency response for $\omega = \pi/2$:

$$H(\pi/2) = C_0 e^0 + C_1 e^{-\pi/2} + C_2 e^{-\pi} + C_3 e^{-3\pi/2} + C_4 e^{-4\pi/2}$$
$$= 1 \cdot 1 + 3 \cdot (-j) + 5 \cdot (-1) + 3 \cdot (j) + 1 \cdot 1 = -3$$

So the magnitude of the frequency response has gone from 13 (at $\omega = 0$) to 3 (at $\omega = \pi/2$). The phase has gone from $0°$ (at $\omega = 0$) to $180°$ (at $\omega = \pi/2$), although generally we are not concerned about the phase response of FIR filters. Just from these two points of the frequency response, we can guess that the filter is probably some type of low-pass filter.

Recall that the magnitude is calculated as follows:

$$\text{Magnitude } Z = X + jY = (X^2 + Y^2)^{1/2} = |Z|$$

Our earlier sample calculation turned out to have only real numbers, but the reason is that the imaginary components of $H(\pi/2)$ canceled out to zero. The magnitude of $H(\pi/2)$ is

$$\text{Magnitude } |-3 + 0j| = 3$$

A computer program can easily evaluate $H(\omega)$ from $-\pi$ to π and plot it for us. Of course, this is almost never done by hand. Figure 5.2 shows a frequency plot of this filter using an FIR filter program.

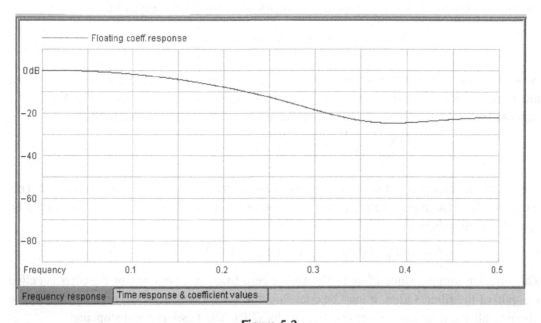

Figure 5.2

This is not the best filter, but it still is a low-pass filter. The frequency axis is normalized to F_s, and the magnitude of the amplitude is plotted on a logarithmic scale, referenced to a passband frequency response of 1.

We can verify our hand calculation was correct. We calculated the magnitude of $|H(\pi/2)|$ at 3 and $|H(0)|$ at 13. The logarithmic difference is

$$20 \log_{10}(3/13) = -12.7 \text{ dB}$$

If you check the frequency response plot shown in the figure, you will see at frequency $F_s/4$ (or 0.25 on the normalized frequency scale), which corresponds to $\pi/2$ in radians, the filter does indeed seem to attenuate the input signal by about 12–13 dB relative to $|H(0)|$. Other filter programs might plot the frequency axis referenced from 0 to π, or from $-\pi$ to π.

5.3 Computing Filter Coefficients

Now suppose you are given a drawing of a frequency response and told to find the coefficients of a digital filter that best matches this response. Basically, you are designing the digital filter. Again, you would use a filter design program to do this, but if you need to do this, it is helpful to have some understanding of what the program is doing. To optimally configure the program options, you should understand the basics of filter design. In this section, we explain a technique known as the Fourier design method. This method requires more math than we have used so far, so if you are a bit rusty, just try to bear with it. Even if you do not follow everything in the rest of the chapter, the ideas should still be very helpful when using a digital filter design program.

The desired frequency response is designated as $D(\omega)$. This frequency response is your design goal. As before, $H(\omega)$ represents the actual filter response based on the number and value of your coefficients. We now define the error, $\xi(\omega)$, as the difference between what we want and what we actually get from a particular filter:

$$\xi(\omega) = D(\omega) - H(\omega)$$

Now all three of these functions are complex; when evaluated, they will have magnitude and phase. We are concerned with the magnitude of the error, not its phase. One simple way to eliminate the phase in the $\xi(\omega)$ is to work with the magnitude squared of $\xi(\omega)$ as in

$$|\xi(\omega)|^2 = \{\text{Real part } \xi(\omega)\}^2 + = \{\text{Imag part } \xi(\omega)\}^2 = \xi(\omega)\xi(\omega)^*$$

where * is the complex conjugate operator (recall from Chapter 2 on complex numbers, magnitude squared is a number multiplied by its conjugate). The squaring of the error function differentially amplifies errors. It makes the error function much more responsive to large errors than smaller errors, usually considered a good thing.

To get the cumulative error, we need to evaluate the magnitude squared error function over the entire frequency response:

$$\text{Error} = \xi = \int_{-\pi}^{\pi} |\xi(\omega)|^2 d\omega$$

The classic method to minimize a function is to evaluate the derivative with respect to the parameter over which we have control. In this case, we will try to evaluate the derivative of ξ with respect to the coefficients, C_i. This will lead us to an expression that allows us to compute the coefficients that result in the minimum error, or minimize the difference between our desired frequency response and the actual frequency response. Because I promised to minimize the math (and many of you probably would not have started reading this otherwise), this derivation is located in Appendix A. If you have trouble with this derivation, do not let it bother you. It is the result that is important anyway.

$$C_i = (1/2\pi) \cdot \int_{-\pi}^{\pi} D(\omega) \, e^{j\omega i} \, d\omega$$

The preceding provides a design equation to compute the filter coefficients that give a response best matching the desired filter response, $D(\omega)$.

Let's try an example. Let $D(\omega)$ be defined as a low-pass filter with cutoff at $\omega = \pi/2$ (see Figure 5.3).

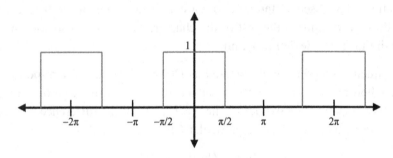

Figure 5.3

We can change the limits on the integral to $-\pi/2$ to $\pi/2$ because $D(\omega)$ is zero in the remainder of the integration interval:

$$C_i = (1/2\pi) \cdot \int_{-\pi}^{\pi} D(\omega) \, e^{j\omega i} \, d\omega = (1/2\pi) \cdot \int_{-\pi/2}^{\pi/2} 1 \cdot e^{j\omega i} \, d\omega$$

From an integration table in a calculus book, we will find

$$\int e^x \, dx = (1/k) \cdot e^{kx}$$

so that we get

$$\int e^{j\omega i}\, d\omega = (1/ji) \cdot e^{j\omega i}$$

The filter coefficients are therefore this expression, evaluated at $-\pi/2$ and $\pi/2$:

$$C_i = (1/2\pi) \cdot \int_{-\pi/2}^{\pi/2} 1 \cdot e^{j\omega i}\, d\omega = (1/2\pi) \cdot (1/ji) \cdot e^{j\omega i}\,|_{-\pi/2}^{\pi/2}$$

Next, we plug in the integral limits:

$$C_i = (1/2\pi) \cdot (1/ji) \cdot [e^{j\pi i/2} - e^{-j\pi i/2}]$$

By using the Euler equation for $e^{j\pi i/2}$ and $e^{-j\pi i/2}$, we find the cosine parts cancel:

$$C_i = (1/2\pi) \cdot (1/ji) \cdot 2j\sin(\pi i/2) = (1/\pi i) \cdot \sin(\pi i/2)$$

This expression gives the ideal response for a digital low-pass filter. The coefficients, which also represent the impulse response, decrease as $1/i$ as the coefficient index i gets larger. It is like a sine wave, with the amplitude gradually diminishing on each side. This function is called the sinc function, also known as $\sin(x)/x$. It is a special function in DSP because it gives an ideal low-pass frequency response. The sinc function is plotted in Figure 5.4, both as a sampled function and as a continuous function.

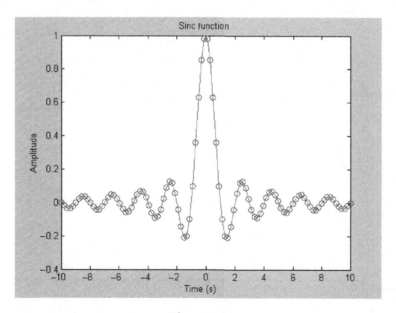

Figure 5.4

Our filter instantly transitions from passband to stopband. But before getting too excited about this filter, we should note that it requires an infinite number of coefficients to realize this frequency response. As we truncate the number of coefficients (which is also the filter's impulse response), we get a progressively sloppier and sloppier transition from passband to stopband as well as less attenuation in the stopband. This can be seen in the figures in the following section. What is important to realize is that to get a sharper filter response requires more coefficients or filter taps, which in turn require more multiplier resources to compute. There will always be a trade-off between quality of filter response and number of filter taps, or coefficients.

5.4 Effect of Number of Taps on Filter Response

A picture is worth a thousand words, or so it is said. The following figures show multiple plots of this filter with the indicated number of coefficients. By inspection, you can see that as the number of coefficients grows, you see actual $|H(\omega)|$ approaching desired $|D(\omega)|$ frequency response.

Filter plots are always given on a logarithmic amplitude scale. This allows us to see passband flatness, as well as see how much rejection, or attenuation, the filter provides to signals in its stopband.

All the following filter and coefficient plots are done using the FIR filter program. These filters can be easily implemented in FPGA or a DSP processor.

Figure 5.5 shows a frequency plot of our ideal low-pass filter. It does not look very ideal. The problem is that it is only 7 coefficients long. Ideally, it should have unlimited coefficients.

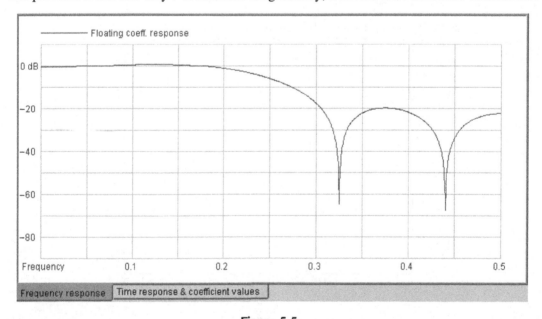

Figure 5.5:
A 7-tap ideal low-pass filter frequency response.

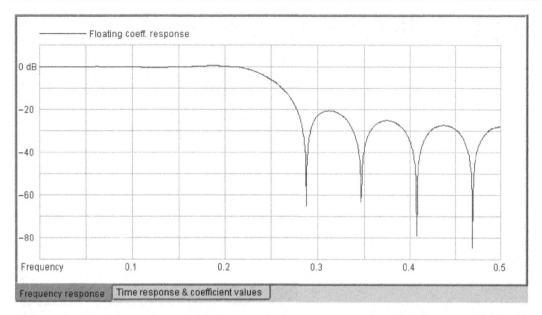

Figure 5.6:
A 15-tap ideal low-pass filter frequency response.

We can see some improvement in Figure 5.6, with about twice the number of taps.

Now this is starting to look like a proper low-pass filter (see Figure 5.7).

You should notice how the transition from passband to stopband gets steeper as the number of taps increases (see Figure 5.8). The stopband rejection also increases.

Figure 5.9 shows a very long filter, with closer to ideal response. Notice how as the number of filter taps grows, the stopband rejection increases (it is doing a better job attenuating unwanted frequencies). For example, using 255 taps, by inspection $|H(\omega = 0.3)| \sim -42$ dB. With 1023 taps, $|H(\omega = 0.3)| \sim -54$ dB. Suppose a signal of frequency $\omega = 0.3$ radians per second, with peak-to-peak amplitude equal to 1, is our input. The 255-tap filter would produce an output with peak-to-peak amplitude of ~ 0.008. The 1023-tap filter would give $4\times$ better rejection, producing an output with peak-to-peak amplitude of ~ 0.002.

Figure 5.10 shows the plot of the 63-tap sinc filter coefficients (this is not a frequency response plot). The sinc shape of the coefficient sequence can be easily seen. Notice that the coefficients fall on the zero crossing of the sinc function at every other coefficient.

There is one small point that might confuse alert readers. Often a filter program will display the filter coefficients with indexes from –N/2 to N/2. In other cases, the same coefficients might be indexed from 0 to N or from 1 to N + 1. Our five-tap example was {C_0, C_1, C_2, C_3,

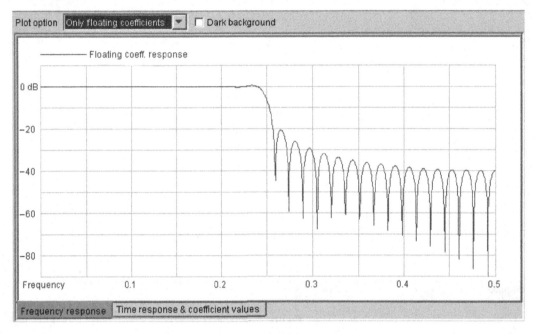

Figure 5.7:
A 63-tap ideal low-pass filter frequency response.

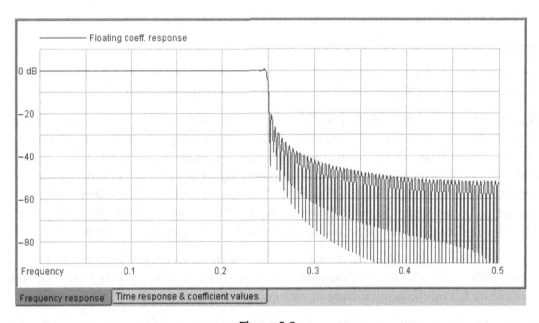

Figure 5.8:
A 255-tap ideal low pass filter frequency response.

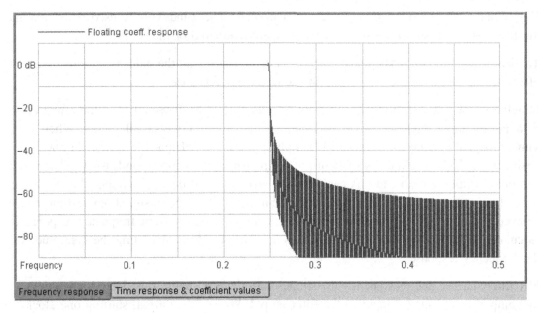

Figure 5.9:
A 1023-tap ideal low-pass filter frequency response.

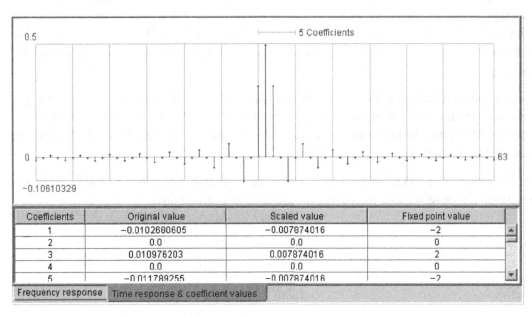

Coefficients	Original value	Scaled value	Fixed point value
1	−0.0102680605	−0.007874016	−2
2	0.0	0.0	0
3	0.010976203	0.007874016	2
4	0.0	0.0	0
5	−0.011789255	−0.007874016	−2

Figure 5.10:
The 63-tap ideal low-pass filter coefficients.

C_4}. What if we use {C_{-2}, C_{-1}, C_0, C_1, C_2} instead? These indexes are used in calculating the filter response, so does it matter how they are numbered or indexed?

The simple answer is no. The reason for this is that changing all the indexes by some constant offset has no effect on the *magnitude* of the frequency response.

Imagine if you put a shift register in front of your FIR filter. It would have the effect of delaying the input sequence by one sample. This is simply a one-sample delay. When a sampled signal is delayed, the result is simply a phase shift. Think for a moment about a cosine wave. If there is a delay of ¼ of the cosine period, this corresponds to 90°. The result is a phase-shifted cosine—in this case, a sine wave signal. It does not change the amplitude or the frequency, only the phase of the signal. Similarly, if the input signal or coefficients are delayed, this is simply a phase shift. As we saw previously, the filter frequency response is calculated as a magnitude function. There is no phase used in calculating the magnitude of the frequency response.

FIR filters have a delay or latency associated with them. Usually, this delay is measured by comparing an impulse input to the filter output. We see the output, starting one clock after the impulse input. This output or filter impulse response spans the length of the filter. So we measure the delay from the impulse input to the largest part of the filter output. In most cases, this is the middle or center tap. For example, with our five-tap sample filter, the delay would be three samples. This is from the input impulse to the largest component of the output (impulse response), which is 5. In general, for an N tap filter, where N is odd, the delay is $N/2 + 1$. When N is even, it is $N/2 + \frac{1}{2}$. Since an FIR filter adds equal delay to all frequencies of the signal, it introduces no phase distortion, and it has a property called linear phase.

This explanation might not be as clear as you would like, but the essence is that since no phase distortion occurs when the signal passes though an FIR filter, we do not need to consider filter phase response in our design process. This is one reason why FIR filters are preferred over other types of digital filters.

The next chapter covers a topic called "windowing," which is a method to optimize FIR filter frequency response without increasing the number of coefficients.

CHAPTER 6

Windowing

In the preceding chapter, we developed a technique for calculating filter coefficients to generate a desired frequency response. Our desired filter response will generate an infinite number of coefficients, and we have to decide how many coefficients to keep, and throw away or truncate the rest. As the previous plots show, as the number of coefficients grows larger, the transition region from passband to stopband grows smaller, and the stopband rejection or attenuation increases.

$$\text{Frequency response} = H(\omega) = \sum_{i=-\infty \text{ to } \infty} C_i e^{-j\omega i}$$

The smaller transition region, or steepness of the transition from passband to stopband, is due to higher frequency components in the frequency response. The higher indexes of "i" in the frequency response correspond to higher frequency complex exponentials. Complex exponentials have a sinusoidal characteristic, so to get a quick response, we must use higher frequency exponentials.

6.1 Truncation of Coefficients

The process of truncating the infinite number of coefficients is called windowing. We can imagine $C_{i=-\infty \text{ to } \infty}$, and multiplying it term by term with a window function, $W_{i=-\infty \text{ to } \infty}$. Let's consider the sample plots from the preceding chapter with our low-pass filter with a cutoff frequency of $\pi/2$:

$$C_i = (1/\pi i) \cdot \sin(\pi i/2) \text{ for } -\infty < i < \infty$$

For our 7-tap filter,

$$W_i = 1 \text{ for } -3 \leq i \leq 3 \text{ and } W_i = 0 \text{ otherwise}$$

Similarly, for our 15-tap filter,

$$W_i = 1 \text{ for } -7 \leq i \leq 7 \text{ and } W_i = 0 \text{ otherwise}$$

In both cases, W_i is a rectangular window. This means that the coefficients are unaltered within the window and are zeroed, or truncated, outside the window. The length of the

Digital Signal Processing 101. DOI: 10.1016/B978-1-85617-921-8.00010-9

rectangular window determines the length of the impulse response of the filter. This is called a rectangular window because the window coefficients W_i are all 1 within the window and 0 outside.

With the rectangular window, we abruptly truncate the impulse response of the filter. Obviously, for realistic filter implementations, we have to limit the impulse response at some point because each tap or coefficient requires a multiplication operation to compute each filter output. But perhaps we can get a more desirable response by reducing the coefficient values gradually at either end of the impulse response before we reach the point of impulse response truncation.

6.2 Tapering of Coefficients

Reducing the coefficient values has led to efforts to develop other window functions besides the default rectangular window. Window design and analysis involve a fair bit of mathematics. But after the rigors of the preceding chapter, you may not mind too much if we skip over this. Actually, many filter designers do not know the details of the various window functions offered by their filter design software but work iteratively instead. That is, designers experiment with moving the frequency cutoff point slightly, and playing the allowable number of taps, the various window options, and sometimes the numerical precision (number of bits) of the input data and coefficients. By observing the computer-generated frequency plots, designers can iterate to find an optimum combination of these parameters to meet application requirements.

Often, the requirements are a certain degree of filter rejection or attenuation at one or more specific frequency points, a maximum amount of ripple or variance in the passband region of the frequency response, and a specified region of the frequency response. (See Figure 6.1.)

Most windows are named after their inventors. They include Hanning, Hamming, von Hann, Kaiser, Blackman, Bartlett, and others. The window coefficients are not equal to 1, as in the rectangular window, but will gradually transition from 1 to 0 in some fashion near the edges of the window. The form of this transition, or tapering off, of the coefficients defines the window properties. Note that this is a different function from filter design, which produces the original and ideal set of coefficients. Windowing is used to avoid the abrupt truncation of the filter coefficient set, required to allow implementation of the filter with a finite number of multiply-add operations.

In general, a window cannot increase the steepness of the transition region, but it can be used to reduce either the passband or stopband ripple in the frequency response. Most filter design programs offer several window options. The following figures show the frequency

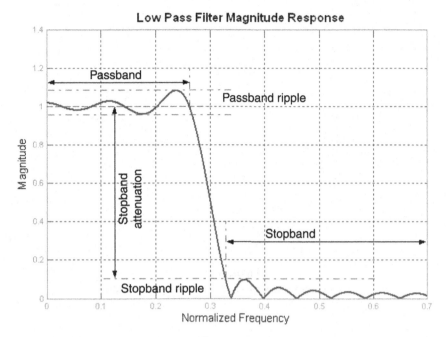

Figure 6.1

responses with different windows for comparison. All the filters shown are for a 61-tap bandpass filter. The major trade-off between the different windows is the width of the transition band and the amount of attenuation in the stopband region of the frequency response. Often, filter designers make trade-offs in transition width, passband ripple, stopband attenuation (including ripple of lobes in the stopband), number of coefficients, and the chosen window to achieve the application requirements.

6.3 Example Coefficient Windows

In Figures 6.2 through 6.5, several windows are shown, with the rectangular window as the baseline. Notice that the rectangular window (Figure 6.2) provides the steepest transition band. However, the stopband sidelobes are very high, reducing the amount of stopband attenuation. The Hanning, Hamming, and Blackman windows provide increasing stopband attenuation, at the expense of a wider transition band. Windowing is supported in all FIR filter design software programs.

These windows have a similar effect on the transition band and sidelobes whether applied to low-pass filters, high-pass filters, or bandpass filters.

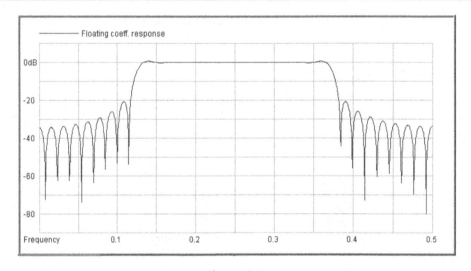

Figure 6.2:
A 61-tap bandpass filter with rectangular window.

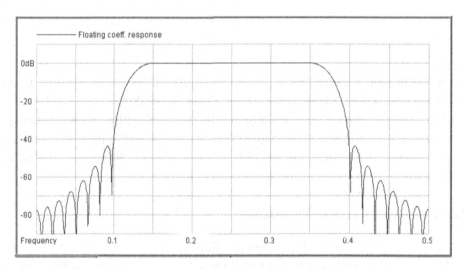

Figure 6.3:
A 61-tap bandpass filter with Hanning window.

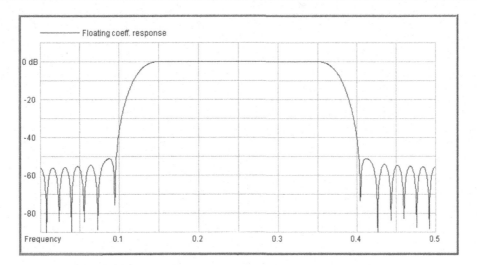

Figure 6.4:
A 61-tap bandpass filter with Hamming window.

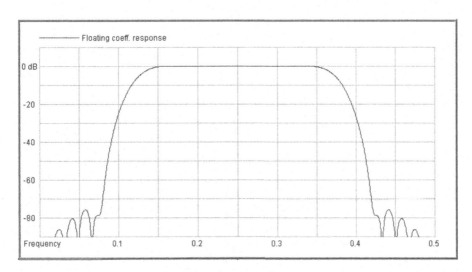

Figure 6.5:
A 61-tap bandpass filter with Blackman window.

CHAPTER 7

Decimation and Interpolation

In this chapter, we discuss decimation and interpolation. Decimation is the process of reducing the sample rate F_s in a signal processing system, and interpolation is the opposite, increasing the sample rate F_s in a signal processing system. These processes are very common in signal processing systems and are nearly always performed using an FIR filter.

First, why are sampling rates changed? The most common reason is to ease the interface of the digital signals to the outside environment. As we saw in previous chapters, signals have a frequency representation, and this frequency representation must be less than the Nyquist frequency, which is defined as $F_s/2$. This sets a lower bound on F_s. The amount of hardware or software processing resources is normally proportional to F_s, so we usually want to keep F_s as small as practical. So while there is no upper bound on F_s, it is usually less than $10\times$ the frequency representation of the signal. A minimum F_s is needed to ensure the highest frequency portion of the signal does not approach the $F_{Nyquist}$ frequency.

In some cases, there are advantages to highly oversampling a signal, where $F_s \gg F_{signal}$. One such example might be when sampling a signal using an analog-to-digital converter (ADC). If F_s is high, then so is $F_{Nyquist}$. Recall that we can accurately represent all signals below $F_{Nyquist}$ in the sampled domain without aliasing occurring. When we make $F_{Nyquist}$ large, any unwanted signals in that frequency space can be eliminated, or at least highly attenuated, by using a digital filter with its passband matching our desired frequency and with its stopband for the rest of the frequency spectrum up to $F_{Nyquist}$. The analog filter needs to filter out frequencies above $F_{Nyquist}$. In this manner, by increasing F_s and therefore $F_{Nyquist}$, we can simplify the requirements for analog filtering prior to the ADC. Once we have successfully filtered out these unwanted signals through both the analog and digital filters, we have no further need for keeping the $F_s \gg F_{signal}$ and should consider lowering F_s to reduce resources required for subsequent processing.

7.1 Decimation

To discuss decimation, or downsampling as it is also called, we need to consider the frequency representation of the signal as well as the time domain. First, let's look at the sampled signal sampled at F_s (see Figure 7.1) and at $F_s{}'$, where $F_s{}' = F_s/2$ (see Figure 7.2). The new sampling rate $F_s{}'$ is the sampling rate after decimating by 2.

Figure 7.3 shows the frequency domain perspective of decimation.

Digital Signal Processing 101. DOI: 10.1016/B978-1-85617-921-8.00011-0

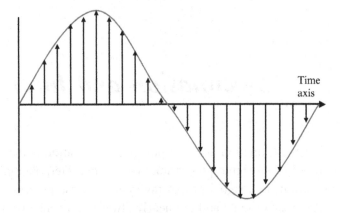

Figure 7.1:
Original sampling rate = Fs.

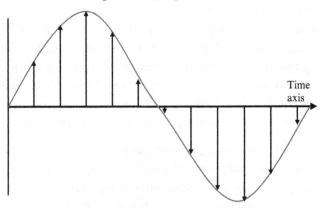

Figure 7.2:
New sampling rate = $F_s' = F_s/2$.

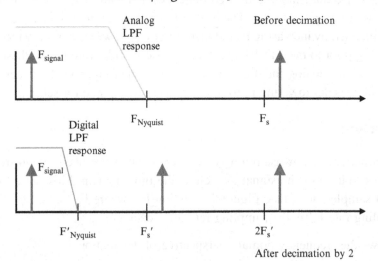

Figure 7.3

As we can see by examining the frequency plots, the signal itself has not changed; only the sampling rate frequency and corresponding Nyquist rate frequency have changed. As long as the new Nyquist frequency is larger than the signal frequency, no aliasing will occur.

The signal images, which are periodic in F_s, and a natural consequence of sampling, are also shown. When we are decimating, the images become periodic in the new sample rate F_s'.

We could decimate by simply throwing away samples. In our example, if we throw away every other sample, the sample rate is reduced by a factor of ½, and the result is as shown in Figure 7.3.

In practice, however, the decimation process usually has a low-pass filter (LPF) incorporated. Let's go back to our ADC example. This could be implemented using the block diagram shown in Figure 7.4. In this example, the analog low-pass filter is responsible for removing all unwanted signals above $F_{Nyquist}$, which would otherwise alias into the frequency region where our signal is. With this analog LPF prior to the ADC, we can then be confident that any signals we see below $F_{Nyquist}$ are legitimate signals, and not aliased versions of some higher-frequency unwanted signal or noise. We must similarly provide an LPF prior to the decimator, to remove any frequency components between $F'_{Nyquist}$ and F_s'; otherwise, any frequency components in this frequency band will alias, or fold back, into the frequency band below our new Nyquist frequency, $F'_{Nyquist}$. The approximate frequency response of the analog and digital LPF is depicted in the previous frequency domain (Figure 7.3).

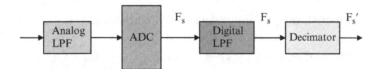

Figure 7.4

Now we can ask an interesting question: Why bother to compute samples at rate F_s in the digital LPF if we are going to discard ½ of them in the decimator? This is a waste of processing resources. Instead, we can build a digital filter that computes only every other sample and therefore accomplishes the function of the decimator (shown separately in our block diagram). And in the process, we find that we need only ½ the multiplier rate to implement the digital LPF.

To see how we might do this, let's consider the FIR block diagram shown in Figure 7.5.

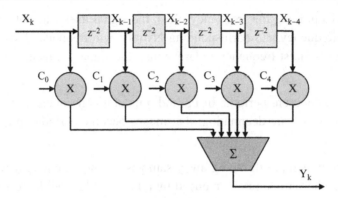

Figure 7.5

In a normal case, at each clock cycle, the x_k data advances into the shift register of the filter structure, and an output y_k is produced. As we did earlier, we can build a sequence of inputs and compute the sequence of outputs. The filter equation is repeated here:

$$y_k = C_0 \cdot x_k + C_1 \cdot x_{k-1} + C_2 \cdot x_{k-2} + C_3 \cdot x_{k-3} + C_4 \cdot x_{k-4}$$

Now suppose that on every even clock cycle (k is even), we operate the filter normally, but on the odd clock cycles (k is odd), we merely shift the input data but disable the operation of multipliers, summation, and update of the output register (this register not explicitly shown). The output register will just hold the previous value, since there has been no update.

Let's look at how to compute a few outputs:

$$y_0 = C_0 \cdot x_0 + C_1 \cdot x_{-1} + C_2 \cdot x_{-2} + C_3 \cdot x_{-3} + C_4 \cdot x_{-4}$$

(y_1 output is not computed; only x_k input data is shifted through);

$$y_2 = C_0 \cdot x_2 + C_1 \cdot x_1 + C_2 \cdot x_0 + C_3 \cdot x_{-1} + C_4 \cdot x_{-2}$$

(y_3 output is not computed; only x_k input data is shifted through);

$$y_4 = C_0 \cdot x_4 + C_1 \cdot x_3 + C_2 \cdot x_2 + C_3 \cdot x_1 + C_4 \cdot x_0$$

and so forth.

The output sequence is the decimated, filtered sequence $\{\ldots y_0, y_2, y_4, \ldots\}$. Only the shift registers operate at the input rate. The rest of the circuitry can be clocked at the output clock rate. If implemented using DSP processors, or if you are clever in your hardware filter design, you can utilize less multipliers by operating them at the faster input speed. This concept is beyond our scope here, but in general, whether you are implementing DSP algorithms in hardware (FPGA or ASIC) or in software (DSP processor), the multipliers can be multiplexed such that it does not matter whether you have a few very fast multipliers or a large number of slow multipliers or anything in between, so long as the cumulative multiply-accumulate capacity is sufficient for the DSP algorithm requirement.

Decimation is limited to integer values. So this concept can be extended to decimate by 3, decimate by 4, decimate by 10, and so forth. In each case, the decimation filter computes only the required samples. The input data is shifted right by the decimation rate between each computation. In our earlier decimate-by-2 example, the input data is shifted right by two places between each output computation (check the x indexes in the sample computations shown previously).

7.2 Interpolation

As we just saw, the sample rate F_s can be decreased by an integer value using a decimation FIR filter. Similarly, the sample rate F_s can be increased by an integer value using a type of FIR filter called an interpolation filter. This is called upsampling, and is the opposite of decimation. As long as the signal frequency content is below the Nyquist frequency at the lowest sampling frequency, we can decimate a signal and then turn around and interpolate it, and recover the same signal.

Interpolation requires that the sample rate be increased by some integer factor. New samples need to be created and inserted between the existing samples. Let us look at the simplest example of interpolation. Let's go back to our sine wave example and interpolate it by a factor of two, as shown in Figure 7.6.

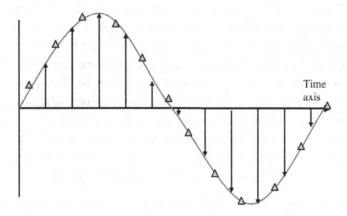

Figure 7.6

This figure shows the original signal at sample rate F_s. Triangles indicate approximately where new samples must be created to interpolate up to sample rate $F_s' = 2 \cdot F_s$. This concept seems straightforward enough. We do not have to worry about aliasing issues because we are doubling both the sample frequency and Nyquist frequency.

The simplest and most intuitive way to interpolate is called linear interpolation. In linear interpolation, we simply draw a straight line between the original samples and calculate the new samples along this line. In our case, if we interpolate by 2, then we need the point located midway along the line between the original points. Linear interpolation, whether by a factor of 2, 3, 4, . . ., is equivalent to drawing a line between all the original points and will look something like that in Figure 7.7.

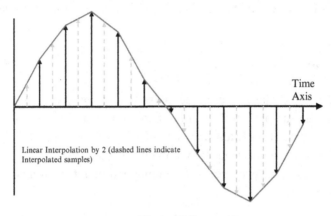

Figure 7.7

The signal at $F_s' = 2 \cdot F_s$ is shown in Figure 7.7 with linear interpolation. Obviously, this doesn't look quite right. But let's consider how we would build an interpolation filter.

An interpolation filter is actually several filters running in parallel, each with the same data input x_k. Each filter computes a different intermediate sample. One filter has a single tap $= 1$, which provides the original signal at the output (this could just be a shift register to provide correct delay). Each of the other filters calculates one of the new samples between the original samples. When we are interpolating by N, there will be N of these filters, including the trivial filter that generates the original samples. For example, when we are interpolating by four, $N = 4$, and there will be four of these filters. Every input sample will produce four output samples, one from each filter, which will be interleaved at the output at a combined rate four times larger than the input rate. This concept of multiple filters creating a single output with a higher rate sequence is called polyphase filtering.

The length or number of taps in each of the N interpolation filters largely determines the quality of the interpolation. With linear interpolation, the number of filter taps is only 2, so the quality of the interpolated signal is rather poor, as can be seen in Figure 7.7. The ideal filter is a low-pass filter with cutoff frequency at $F_{Nyquist}$ of the original signal sampling rate. As we learned previously, an ideal filter has an infinite number of coefficients. We will have to compromise at less than infinity, and so each individual filter will have M taps.

Shown in Figure 7.8 is an interpolate-by-4 (N = 4) polyphase filter, with 5 taps (M = 5) used for each phase. The input data stream x_k is sampled at F_s, and the serialized output data stream y_m is sampled at $F_s' = N \cdot F_s$. Coefficient representations A_n, B_n, C_n are used to indicate that each phase of the interpolation filter may have a different set of coefficients.

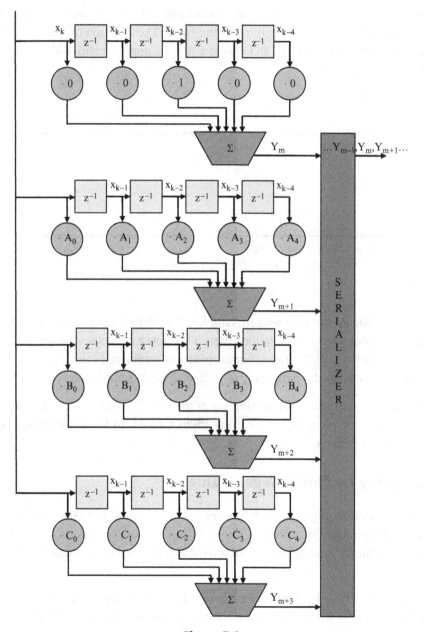

Figure 7.8

You should note how the first filter could be eliminated and replaced by a shift register. All the taps except the center one multiply by zero, so the multiplier and adder logic is not required. The original input sample simply passes through with some delay.

Interpolating polyphase filters is a little tricky, so we also show what is happening in the time or sample domain. Figure 7.9 shows the input and output sequences, which are time aligned for clarity. The output data rate is four times the input, and as with all FIR filters, there is some processing delay through the filter. The function of the serializer is to input N samples in parallel at rate F_s and output a serial sample stream at F_s', which is the new interpolated sample rate.

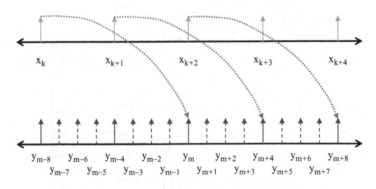

Figure 7.9

The dashed lines indicate the interpolated samples. The dotted lines show the delay of the original samples through the interpolation filter. This delay depends on filter design but is generally equal to $(M - 1)/2$ input samples, plus any additional register pipeline delays. In our example, $M = 5$, so the delay is $(5 - 1)/2 = 2$ input sample intervals.

Note that the "m" output index is increasing faster than the "k" input index. The interpolation filter must produce N output samples for every input sample, so the output index "m" needs to run N times faster than "k."

7.3 Resampling by Non-Integer Value

Suppose that you need to align the sample rates between sets of digital circuitry running at different sampling rates. For example, the sampling rate of the first circuit is 3 MSPS, and the second circuit has a sampling rate of 2 MSPS. You need to decimate, or downsample, by 2/3.

You can achieve this effect by a combination of interpolating and decimating. In this case, you would interpolate by 2 and then decimate by 3, as shown in Figure 7.10.

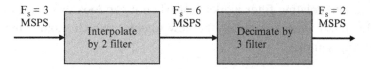

Figure 7.10

Next, a frequency domain representation of both interpolation and decimation steps is shown in Figure 7.11.

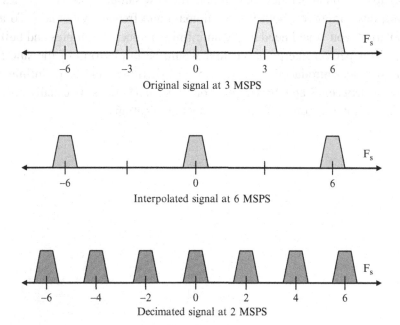

Figure 7.11

Both the interpolation and decimation filters incorporate a low-pass filtering function. The reason for this low-pass filter, however, is quite different for each case. For decimation, the low-pass filter serves to eliminate high-frequency components in the spectrum. If these components were not filtered out, they would alias when the reduction in sample rate is performed.

For interpolation, the low-pass filter serves to provide a "smoothing" function when calculating the new samples so that a smooth curve results when the new samples are inserted between the original samples. A longer interpolating filter (more taps = larger M) will use a weighted calculation of a larger number of the adjacent samples to determine the new samples. An analogy might be when a person is driving a car on a winding road. If you look only a few feet in front of the car, you cannot take the curves smoothly. Since you are not looking ahead, you cannot anticipate the direction and rate of curves and smoothly adjust your driving. The interpolating filter works best when it can look at samples on both sides (or in front and behind) when computing the new samples, which should smoothly fit in between the existing samples.

When M = 2, which is linear interpolation, only the two adjacent samples are used, and the filter computes the sample that lies on a straight line between the two points. As we saw in the example, this does not give a very smooth response. This lack of smooth response can also create some higher-frequency components in the signal spectrum. If M = 4, then the filter uses 4 samples, 2 on either side, to compute the new samples. As M becomes larger, the interpolated response improves, both in time domain and frequency domain. To achieve perfect interpolation, you would need to use an infinite number of samples and build a perfect low-pass filter. This perfect interpolation filter would be in the form of the sinc function, also known as sin(x)/x, introduced in Chapter 5 on FIR filters, and be of infinite length. In practice, using between 8 and 16 samples, or filter coefficients, is usually enough to a reasonable job of interpolating a signal for most applications.

CHAPTER 8

Infinite Impulse Response (IIR) Filters

This chapter discusses infinite impulse response (IIR) filters. The IIR filter is unfortunately a much more complex topic than the finite impulse response (FIR) filter, and due to its nonlinear behavior, it is very difficult to analyze. It is also more complex to implement, due to the feedback, though it typically does require fewer multipliers. For these reasons, IIR filters are used much less often than FIR filters. On the plus side, since they are not commonly used, understanding IIR filters is not essential to the fundamental concepts of DSP. The IIR filter design technique is usually considered a bit of a specialty in the DSP world, so do not feel that you need to master this topic.

All the popular IIR filter designs are based on analog filter circuits. The nature of analog components—capacitors, inductors, opamps—tends to be naturally suited to recursive filter designs. By contrast, FIR filters are naturally implemented digitally. This chapter introduces IIR filters and then focuses primarily on how to convert an analog IIR filter into a sampled digital implementation.

Mathematics for IIR filters tends to be more daunting. If the math proves too much, note that the discussion in this chapter is not necessary to continue on to the other topics in the remainder of this book. Plus, the following chapter on digital modulation is really interesting; you would not want to miss it.

Most digital IIR filter designs are derived from analog filter designs. Because analog filters were around long before digital filters existed, this provides for many types of filters.

The basic design procedure is to take an analog filter design and convert it to a digital IIR filter implementation. Since we do not have time or space to go through analog filter fundamentals, some material presented here may be difficult if you do not have any familiarity with analog filter design techniques. Now, enough of the disclaimers; let's begin.

An IIR filter is basically an FIR filter with feedback added. The FIR filter takes a stream of input data and multiplies it by a series of coefficients to produce each output. The IIR filter also does this, but in addition, it feeds the output data stream back through another series of multipliers and coefficients. This structure is shown in the diagram in Figure 8.1.

Digital Signal Processing 101. DOI: 10.1016/B978-1-85617-921-8.00012-2

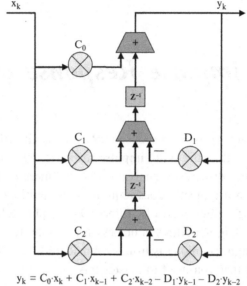

$$y_k = C_0 \cdot x_k + C_1 \cdot x_{k-1} + C_2 \cdot x_{k-2} - D_1 \cdot y_{k-1} - D_2 \cdot y_{k-2}$$

Figure 8.1

This feedback eliminates many of the linear properties of the FIR filter and makes the IIR filter much more difficult to analyze. It can also create some undesired behavior, depending on the choice of coefficients, where the impulse response may have an infinite duration or even an infinite magnitude.

8.1 IIR and FIR Filter Characteristic Comparison

Before we get too far into IIR filters, it is useful to compare IIR and FIR filters. Following is a summary of advantages and disadvantages of IIR and FIR digital filters:

- FIR filters have a linear phase, meaning that no phase distortion of the signal occurs. IIR filters always cause some phase distortion,* so filter designers need to consider the phase as well as magnitude response.
- FIR filters are always stable and have a finite length impulse response. IIR filters generally have an infinite length impulse response and may have infinite magnitude output (become unstable) under some conditions.
- FIR filters can be designed with a specified amount of quantization noise (remember quantization from Chapter 1 on numerical representation), which can be made as small as necessary. This is not the case with IIR filters.

---

* Phase distortion happens when the phase response of the filter changes nonlinearly across the filter's passband frequency response. Recall that filters are complex; they can affect both the magnitude and phase of the signal.

- FIR filters can be implemented efficiently in multirate systems or systems that have decimation or interpolation steps.
- IIR filters are very sensitive to coefficient values and numerical precision in designs that require a sharp cutoff frequency response.
- IIR filters are a more natural digital form to replace existing analog filters.
- An IIR filter can provide a much sharper cutoff frequency response compared to the same order FIR filter. In other words, for sharp response in an FIR filter, more resources (multipliers and adders) are required than in an IIR filter.

Based on these reasons, you might wonder why anyone would want to use IIR filters. Well, aside from the last item, which lists a key advantage of IIR filters, the reason is that IIR filters can best approximate the performance of many analog filter responses. Sometimes when a system with analog filtering is being upgraded to digital implementation, it is important to preserve its performance characteristics, especially in the phase domain. An example of this might be professional audio equipment.

Intuitively, the reason IIR filters can have a much more rapid frequency response is that their frequency response is determined by both zeros and poles. Zeros are caused by cancellations. Remember the FIR discussion, with coefficients of $+1$, -1, $+1$, -1, and so on? This filter causes cancellations at high frequencies, and this can be described as zero at a specific frequency. Filter response is determined by canceling various frequencies.

Poles, on the other hand, involve feedback and result in an effect similar to division by zero at specific frequencies. Actually, it is not exactly divided by zero because this would produce an infinite response, but the idea is to divide by a small number at certain frequencies, which can produce a large output in the signal in that frequency region. Think of a rope representing frequency response. The poles would hold up the rope at specific points, and the zeros would be like a lead weight holding it to the ground at other points. This is analogous to how poles and zeros can act on the frequency response.

The additional flexibility of using poles with zeros, while resulting in some unstable filter responses if we are not careful (imagine a pole that is very high or even infinite), can also provide for more rapid changes in the frequency response. This makes IIR filters more sensitive and complicated than FIR filters.

In this introductory chapter, we do not try to explain the design of IIR filters using pole and zero placement. There are whole books on this topic, and it is too big for this introductory treatment here. Rather, we show how to take popular analog filter designs, with a defined pole and zero arrangement and thus frequency response, and convert to a digital implementation because this is a more common task for DSP system designers.

Even a basic understanding of IIR filters requires some mathematics. The reason is that analog filters are analyzed using something called the Laplace transform. Digital filters are analyzed using something called the z-transform. Because of their simplicity, we managed to avoid the z-transform when discussing FIR filters, but we do not have that option here. Refer to Appendices C and D at the end of book introducing Laplace and z-transforms if they are new to you.

8.2 Bilinear Transform

The normal design procedure for IIR filters is to specify the filter response and design an analog filter using analog filter techniques (using the Laplace transform). Alternately, you might be given an analog filter design and be asked to convert it to a digital implementation. The analog filter design is based on the location of poles and zeros in the s-plane. A digital filter response can be characterized by using the z-transform. The equivalent pole and zero domain for digital filters is called the z-plane. The idea is to map the s-plane poles and zeros to the z-plane. The mapping technique most often used between the s- and z-domains is called the bilinear transform. There are alternative techniques, but we do not cover them here. Further, only a rudimentary coverage is attempted here because the topics involved are fairly mathematical.

Again, note that both the Laplace and z-transforms are reviewed in Appendices C and D at the end of this book. If you have not looked at them already, you might want to spend a little time going through the Laplace and z-transform appendices.

We use $\acute{\omega}$ for the s-plane to distinguish between ω of the sampled domain z-plane. Similarly, we define the frequency response of the analog filter as H_s and the frequency response of the digital filter as H_z.

In analog filter design, the frequency response of an analog filter is defined by setting $s = j\acute{\omega}$ and evaluating for $-\infty < \acute{\omega} < \infty$. This corresponds to the imaginary axis of the s-plane.

The frequency response of a digital filter is defined by setting $z = e^{j\omega}$ and evaluating for $-\pi < \omega < \pi$. This corresponds to the unit circle of the z-plane.

To go between these two domains, we need a mapping function from the s-plane to the z-plane. Then we can map the zeros and poles across the two domains.

This task is performed by replacing s in the expression for H_s

$$s = 2 \cdot (1 - z^{-1})/(T \cdot (1 + z^{-1}))$$

where $T = 1 / F_s$ (T = time between sampling interval of the digital filter in seconds).

Let's go through an example of converting an analog filter to a digital IIR filter. There are a great many analog filter types. Here, we discuss only one because we do not want to focus on analog filters. A very common analog filter is the Butterworth filter. It has the characteristic of not having ripples in the passband or stopband.

We are going take a third-order Butterworth analog filter and convert it to an IIR digital filter using the bilinear transformation technique. Many analog filters are known simply by their pole and zero locations (since this is another way of defining frequency response). We can eliminate quite a bit of algebra by using the previously described relationship between s- and z-domains to come up with a relationship between poles and zeros in the s- and z-domains. These derivations are not discussed here, only the result.

The digital filter has poles and zeros at the following locations:

$$\text{z-pole}_i = (2 + \text{s-pole}_i \cdot T)/(2 - \text{s-pole}_i \cdot T)$$

$$\text{z-zero}_i = (2 + \text{s-zero}_i \cdot T)/(2 - \text{s-zero}_i \cdot T)$$

Let's set the cutoff frequency of our third-order Butterworth filter to 100 Hz, or 628 rad/s $(100 \cdot 2 \cdot \pi)$. For a third-order Butterworth filter, the three pole locations in the s-plane are located at

$$\text{s-pole}_1 = 628 \text{ angle } (120°) = -314 + j544$$

$$\text{s-pole}_2 = 628 \text{ angle } (180°) = -628$$

$$\text{s-pole}_3 = 628 \text{ angle } (240°) = -314 - j544$$

Now we set our digital sampling frequency at 1000 samples per second, so $T = 0.001$. The s-domain poles map to the following poles in the z-plane:

$$\text{z-pole}_1 = [2 + (-314 + j544) \cdot 0.001]/[2 - (-314 + j544) \cdot 0.001]$$
$$= 0.745 \text{ angle } (72.56°)$$

$$\text{z-pole}_2 = [2 + (-628) \cdot 0.001]/[2 - (-628) \cdot 0.001] = 0.523$$

$$\text{z-pole}_3 = [2 + (-314 - j544) \cdot 0.001]/[2 - (-314 - j544) \cdot 0.001]$$
$$= 0.745 \text{ angle } (-72.56°)$$

The Butterworth filter has three zeros located at infinity. They can be evaluated as

$$\text{z-pole}_{1,2,3} = [2 + \infty \cdot 0.001]/[2 - \infty \cdot 0.001] = \infty/-\infty = -1$$

Now that we have the poles and zeros on the z-plane, we can determine the z-transform of the digital IIR filter approximating the response of the analog Butterworth filter:

$$H(z) = \left[\prod_{i=0 \text{ to } M}(z\_z - \text{zero}_i)\right] / \prod_{i=0 \text{ to } N}(z\_z - \text{pole}_i)]$$

where M = number of zeros and N = number of poles. In our example, M = N = 3. With a bit of algebra, we can multiply this whole mess out to get the more familiar form:

$$H(z) = \left(\sum_{i=0 \text{ to } N} C_i \cdot z^{-i}\right) / \left(1 - \sum_{i=1 \text{ to } M} D_i \cdot z^{-i}\right)$$

In this form, we can pick off our coefficients needed to implement the IIR filter.

Multiplying the numerator and denominator components, we get

$$H(z) = \frac{z^3 + 3z^2 + 3z + 1}{z^3 - 0.970z^2 + 0.108z - 0.290} = \frac{1 + 3z^{-1} + 3z^{-2} + z^{-3}}{1 - (0.970z^{-1} - 0.108z^{-2} + 0.290z^{-3})}$$

From inspection, we can see

$$C_0 = 1 \qquad C_1 = 3 \qquad C_2 = 3 \qquad C_3 = 1$$

and

$$D_1 = 0.970 \qquad D_2 = -0.108 \qquad D_3 = 0.290$$

These coefficients apply to the IIR filter structure depicted earlier in the chapter (although this is a third-order filter, the sample diagram is second order).

8.3 Frequency Prewarping

We still have another point to consider. Our technique is to take the pole and zero locations (in the s-plane) of an analog filter and to map them to pole and zero locations (in the z-plane) of an IIR digital filter.

The problem is that this relationship is not linear. For example, the y-axis of the s-plane, which is infinite in length, is mapped to the unit circle in the z-plane. This relationship is

$$\acute{\omega}_{s\text{-plane}} = (2/T) \cdot \tan(\omega/2) \text{ and}$$

$$\omega_{z\text{-plane}} = (2/T) \cdot \tan^{-1}(\acute{\omega}T/2), \text{where } T = 1/F_s$$

This nonlinear mapping causes filter response distortion, particularly at higher frequencies. A method to mitigate this is to prewarp the analog filter. The whole analog filter is not prewarped; instead, key breakpoints in the analog filter response are prewarped. This prewarping, in essence, stretches the analog filter in frequency, so that when it is compressed by the bilinear transform, the prewarped breakpoint(s) is(are) mapped correctly. This ensures that transition points between passband and stopband are accurately converted by the bilinear transform.

The following table shows the distortion caused by mapping between the s- and z-domains. For this table, we set T = 1. Note how this distortion or warping increases with frequency.

Table 8.1: Mapping between s-domain and z-domain

$\acute{\omega}$, Analog Frequency (s-plane)	ω, Digital Frequency (z-plane)
$0.0 \cdot \pi$	$0.0 \cdot \pi$
$0.1 \cdot \pi$	$0.0992 \cdot \pi$
$0.2 \cdot \pi$	$0.1938 \cdot \pi$
$0.3 \cdot \pi$	$0.2804 \cdot \pi$
$0.4 \cdot \pi$	$0.3571 \cdot \pi$
$0.5 \cdot \pi$	$0.4238 \cdot \pi$
$0.6 \cdot \pi$	$0.4812 \cdot \pi$
$0.7 \cdot \pi$	$0.5302 \cdot \pi$
$0.8 \cdot \pi$	$0.5702 \cdot \pi$
$0.9 \cdot \pi$	$0.6081 \cdot \pi$
$1.0 \cdot \pi$	$0.6391 \cdot \pi$

Let's consider a simple example. Say we have an analog low-pass filter, with a 3 dB breakpoint at 100 Hz (a 3 dB breakpoint is the point in the transition band where filter response is 3 dB lower than the passband response). We want to implement this filter with a digital IIR filter, which has an $F_s = 1000$ Hz.

The digital breakpoint frequency should be $(100/1000) \cdot 2\pi = 0.2 \cdot \pi$. The warping will cause this breakpoint to occur at digital frequency $0.1938 \cdot \pi$ instead.

The analog filter breakpoint needs to be moved, or prewarped, to compensate for this. Substituting into $\acute{\omega}$ (s-plane) $= (2/T) \cdot \tan(\omega/2)$, we find

$$\acute{\omega} = (2/0.001) \cdot \tan(0.2 \cdot \pi/2) = 649.8 \text{ rad/s} = 103.4 \text{ Hz}.$$

We should design the analog filter with a 3 dB point at 103.4 Hz, rather than 100 Hz prior to converting to a digital IIR filter using the bilinear transform to map the pole/zero locations.

To illustrate the effect as the frequency rises, let's revise the problem to building a digital filter to replace an analog filter with a 3 dB breakpoint at 250 Hz. We still can use $F_s = 1000$ Hz.

The digital frequency of the breakpoint is $\omega = (250/1000) \cdot 2\pi = 0.5 \cdot \pi$.

The analog frequency of the breakpoint needs to be moved from 250 Hz to prewarp the filter. The new analog filter breakpoint is found to be

$$\acute{\omega} = (2/0.001) \cdot \tan(0.5 \cdot \pi/2) = 2000 \text{ rad/s} = 318.3 \text{ Hz}$$

We can see the importance of prewarping the analog filter prior to applying the bilinear transform as we move our breakpoint closer to the Nyquist frequency of the digital filter (in this example, equal to 500 Hz). If there are multiple transition points, they can all be prewarped, and the analog filter modified to meet these new transition points.

If you do not have much analog filter design experience, you may be wondering how to modify the analog filter to find prewarped analog poles and zeros that will then be mapped to the digital domain. This is a valid concern but unfortunately beyond the scope of this chapter. The focus of our discussion is how to learn the basics of converting an analog filter to a digital IIR filter. Since analog filter design is a complex and mathematical subject, this issue is just too much to try to cover here. However, as you might expect, there are software programs you can use to perform analog filter design and even convert them to a digital IIR design.

CHAPTER 9

Complex Modulation and Demodulation

We are going to take an unusual approach here. The normal explanations on modulation and demodulation are heavily based on mathematics and equations. In this chapter, we try to take an almost entirely intuitive approach, based on examples. We do not attempt to establish any mathematical foundation or to calculate performance.

Modulation is the process of taking information bits and mapping them to symbols. The sequence of symbols is then filtered to produce a baseband waveform, with the desired spectral properties. This baseband waveform or signal is then upconverted to a carrier frequency, which can be transmitted over the air, through coaxial cable, or through fiber or some other medium. The key idea here is the concept of a symbol.

9.1 Modulation Constellations

We are going to use a common modulation method, known as quadrature phase shift keying (QPSK), as our first example. Don't let the technical-sounding names of these different modulations confuse you. We describe what these names mean later. With QPSK, every two input bits map to one of four symbols, as shown in Figure 9.1.

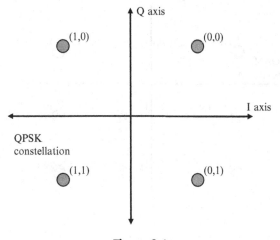

Figure 9.1

Digital Signal Processing 101. DOI: 10.1016/B978-1-85617-921-8.00013-4

The bitstream of zeros and ones input to the modulator is converted into a stream of symbols. Each symbol is represented as the coordinates of a location in the I-Q plane. In QPSK, there are four possible symbols, arranged as shown. Since there are four symbols, the input data is arranged as groups of 2 bits, which are mapped to the appropriate symbol. The arrangement of symbols in the I-Q plane is also called the constellation.

Table 9.1: QPSK constellation mapping

Input Bit Pair	Symbol Location on Complex Plane	I Value	Q Value	Symbol Value (Same as Location on Complex Plane)
0, 0	1 + j	1	1	I+jQ => 1 + j
0, 1	1 - j	1	-1	I+jQ => 1 - j
1, 0	-1 + j	-1	1	I+jQ => -1 + j
1, 1	-1 - j	-1	-1	I+jQ => -1 - j

Another common modulation scheme is known as 16-quadrature amplitude modulation (16-QAM), which has 16 symbols, arranged as shown in Figure 9.2. Again, do not worry about the name of the modulation. Since we have 16 possible symbols, each symbol maps to 4 bits. To put it another way, in QPSK, each symbol carries 2 bits of information, while in 16-QAM, each symbol carries 4 bits of information.

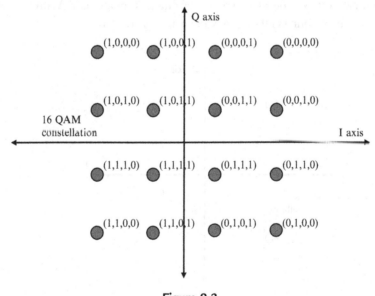

Figure 9.2

We can easily see that the 16-QAM is the more efficient modulation method. In this chapter, we are going to pick a convenient symbol rate and reference our discussion to this symbol rate. But this is an arbitrary choice on my part, and systems are designed with symbol rates ranging from a few kilohertz to hundreds of megahertz. We will decide to transmit symbols at a rate of 1 MHz, or 1 million symbols per second (MSPS). Then our system, if using the 16-QAM modulation, is able to send 4 megabits per second (Mbps). If instead QPSK is used in this system, it is able to send only 2 Mbps. We could also use an even more efficient constellation, 64-QAM. Since there are 64 possible symbols, arranged as 8 rows of 8 symbols each, then each symbol carries 6 bits of information and supports a data rate of 6 Mbps. This is shown in the following table for a few sample constellation types.

Table 9.2: Modulation data rates

Modulation Type	Possible Number of Symbols	Bits per Symbol	Transmitted Bit Rate
QPSK	$4 = 2^2$	2	2 * symbol rate
8 PSK	$8 = 2^3$	3	3 * symbol rate
16-QAM	$16 = 2^4$	4	4 * symbol rate
64-QAM	$64 = 2^6$	6	6 * symbol rate
256-QAM	$256 = 2^8$	8	8 * symbol rate

The frequency bandwidth is determined mainly by the symbol rate. A QPSK signal at 1 MSPS requires about the same bandwidth as a 16-QAM signal at 1 MSPS. Notice the 16-QAM modulator is able to send twice the data within this bandwidth, compared to the QPSK modulator. There is a trade-off, however. As we increase the number of symbols, it becomes more and more difficult for the receiver to detect which symbol was sent. If the receiver needs to choose among 16 possible symbols that could have been transmitted rather than choose from among 4 possibilities, it is more likely to make errors. The level of errors depends on the noise and interference present in the receive signal, the strength of the receive signal, and how many possible symbols the receiver must select from. In cellular systems, there are often high levels of interfering noise or weak signals due to buildings or other objects blocking the transmission path. In this situation, it is often preferable to use a simple constellation, such as QPSK. Even with a weak signal, the receiver can usually make the correct choice of four possible symbols. Other systems, such as microwave radio systems, usually have directional receive and transmit antennas facing each other from building roofs or mountaintops. Consequently, the interfering noise level is usually very low, and complex constellations such as 64-QAM or 256-QAM can be used. This assumes the receiver is able to make the correct choice from among 64 symbols, which allows three times more bits to be encoded into each symbol, resulting in a $3\times$ higher data rate. Recently, sophisticated communication systems such as LTE and WiMax have been introduced, allowing the transmitter to dynamically switch between constellation types depending on quality of reception the receiver experiences.

9.2 Modulated Signal Bandwidth

Now that we have discussed what a constellation is, we still need to discuss some of the steps in taking this set of constellation points and transmitting them over some medium to a receiver. Let's take a look at a QPSK constellation, with a transmission rate of 1 MSPS. The baseband signal is two dimensional, so it must be represented with two orthogonal components, which are, by convention, denoted I and Q.

Let us consider a sequence of 5 QPSK symbols, at time t = 1, 2, 3, 4, and 5, respectively. First, let's look at the sequence in the two-dimensional complex constellation plane (see Figure 9.3). It appears as a signal trajectory moving from one constellation point to another over time.

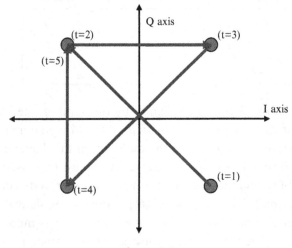

Figure 9.3

We can also look at the I and Q signals individually, plotted against time in Figure 9.3. We can take this two-dimensional signal and plot each component separately (see Figure 9.4). In actuality, the I and Q baseband signals are generated as two separate signals and later combined together with a carrier frequency to form a single passband signal. This topic is discussed in Chapter 11.

Notice the sharp transitions of the I and Q signals at each symbol. Intuitively, we know that a sharp transition requires the signal to have high-frequency content. A signal that is of low frequency can change only slowly and smoothly.

The high-frequency content of these I and Q signals can cause problems, because in most systems, it is important to minimize the frequency content, or bandwidth, of the signal. Remember the early discussion on frequency response, where a low-pass

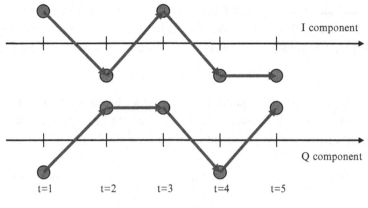

Figure 9.4

filter removes fast transitions or changes in a signal (or eliminates the high-frequency components of that signal). If the frequency response of the signal is reduced, this is the same as reducing its bandwidth. The smaller the bandwidth of the signal, the more signal channels and, therefore, capacity can be packed into a given amount of frequency spectrum. Thus, the channel bandwidth is often carefully controlled.

A simple example is FM radio. Each station, or channel, is located 200 kHz from its neighbor. This means that each station has a 200 kHz spectrum, or frequency response, it can occupy. The station on 101.5 is actually transmitting with a center frequency of 101.5 MHz. The channels on either side transmit with center frequencies of 101.3 and 101.7 MHz. Therefore, it is important to restrict the bandwidth of each FM station to within ±100 kHz, which ensures it does not overlap, or interfere, with neighboring stations. As we know by now, one way to restrict the bandwidth of a signal is to use a filter.

In this discussion, we assume that a given signal's frequency response, or spectrum, can be moved up or down the frequency axis at will. This is true, and is called upconversion or downconversion, and will be discussed in Chapter 11.

9.3 Pulse-Shaping Filter

To accomplish this frequency limiting of the modulated signal, we need to pass the I and Q signals through a low-pass filter. This filter is often called a pulse-shaping filter, and it determines the bandwidth of the modulated signal. But it is not quite that simple. We need to consider what the filter does to the signal in the time domain as well.

Suppose that we use an ideal low-pass filter. Let's use our example where symbols are generated at a rate R of 1 MSPS. The period T is the symbol duration, and equal to 1 μs in our example. The relationship between the rate R and symbol period T is

$$R = 1/T \text{ and } T = 1/R$$

If we alternate with positive and negative I and Q values at each sample interval (this is the worst case in terms of high-frequency content), the rate of change is 500 kHz. So we start with a low-pass filter with a passband of 500 kHz (see Figure 9.5).

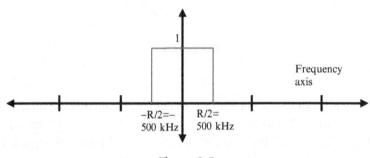

Figure 9.5

This filter has the sinc impulse or time response. The impulse response is shown above in Figure 9.5. It has zero crossings at intervals of T seconds, and decays slowly. A very long filter is needed to approximate the sinc response. The impulse responses of the symbols immediately preceding and following the center symbol are shown in Figure 9.6. The actual transmitted signal will be the sum of all the symbols' impulse response (we just show three symbols here). If the I or Q sample has a negative value for a particular symbol, then the impulse response for that symbol will be inverted from what is shown in Figure 9.6.

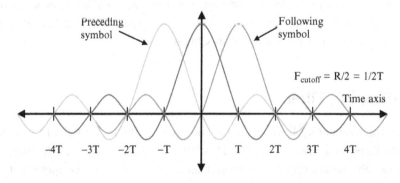

Figure 9.6

Think for a moment about the job of the receiver. The receiver is sampling the signal at T intervals in time to detect the symbol values. It must sample at the T intervals shown on the time axis in Figure 9.6 (leave aside for now the question of how the receiver knows exactly where to sample). At time t = –T, the receiver samples the first symbol. Notice how the two later symbols have zero crossings at t = –T, and so have no contribution at this instant. At t = 0, the receiver samples the value of the second symbol. Again, the other symbols, such as the first and third adjacent symbols, have zero crossings at t = 0 and have no contribution. If we were to reduce the bandwidth of the filter to less than 500 kHz (R/2), then in the frequency domain, these pulses would widen (remember that the narrower the frequency spectrum, the longer the time response, and vice versa). This scenario is shown in Figure 9.7 if the F_{cutoff} of the pulse-shaping filter is narrowed to 250 kHz, or R/4.

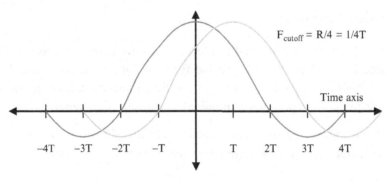

F_{cutoff} = R/4 = 1/4T

Time axis

–4T   –3T   –2T   –T   T   2T   3T   4T

Figure 9.7

In this case, notice how at time t = 0, the receiver samples contributions from all three pulses. At each sampling point of t equal to . . .–3T, –2T, –T, 0, T, 2T, . . ., the signal is going to have contributions from many nearby symbols, preventing detection of any specific symbol. This phenomenon is known as intersymbol interference (ISI) and shows that to transmit symbols at a rate R, we need to have at least R Hz (or 1/T Hz) in the passband frequency spectrum. At baseband, the equivalent two-dimensional (complex) spectrum is from –R/2 to +R/2 Hz to avoid creating ISI. Therefore, to transmit a 1 MSPS signal over the air, at least 1 MHz of RF frequency spectrum is required. The baseband filters need a cutoff frequency of at least 500 kHz.

Notice that the frequency spectrum or bandwidth required depends on the symbol rate, *not* the bit rate. We can have a much higher bit rate, depending on the constellation type used. For example, each 256-QAM symbol carries 8 bits of information, while a QPSK symbol carries only 2 bits of information. But if they both have a common symbol rate, both constellations require the same bandwidth.

We still have two problems, however. One is that the sinc impulse response decays very slowly and so takes a long filter (many multipliers) to implement. The second is that although the response of the other symbols does go to zero at the sampling time when $t = N \cdot T$, where N is any integer, we can see visually that if our receiver samples just a little bit to either side, the adjacent symbols will contribute. This makes the receiver symbol detection performance very sensitive to our sampling timing.

Ideally, we want an impulse response that still goes to zero at intervals of T but decays faster and has lower amplitude lobes, or tails. This way, if we sample a bit to one side of the ideal sampling point, the lower amplitude tails make the unwanted contribution of the neighboring symbols smaller. By making the impulse response decay faster, we can reduce the number of taps and, therefore, multipliers required to implement the pulse-shaping filter.

9.4 Raised Cosine Filter

There is a type of filter commonly used to meet the requirements described in the preceding section. It is called the raised cosine filter, and it has an adjustable bandwidth, controlled by the "roll-off" factor. The trade-off is that as bandwidth of the signal becomes a bit wider, more frequency spectrum is required to transmit the signal.

The following table summarizes the raised cosine response shown in Figure 9.8 for different roll-off factors. These labels are also used in Figures 9.10 and 9.11.

Table 9.3: Roll-off factor characteristics

Roll-off Factor	Label	Comments*
0.10	A	Requires long impulse response (high multiplier resources), has small frequency excess bandwidth of 10%
0.25	B	A commonly used roll-off factor, excess bandwidth of 25%
0.50	C	A commonly used roll-off factor, excess bandwidth of 50%
1.00	D	Excess bandwidth of 100%, never used in practice

*Excess bandwidth refers to percentage of additional bandwidth required compared to ideal low-pass filter.

In Figure 9.8, the frequency response of the raised cosine filter is shown. So that we can better see the passband shape, it is plotted linearly rather than logarithmically (dB). It has a cutoff frequency of 500 kHz, the same as our ideal low-pass filter. A raised cosine filter response is wider than the ideal low-pass filter, due to the transition band. This excess frequency bandwidth is controlled by a parameter called the "roll-off" factor. The frequency response is plotted for several different roll-off factors. As the roll-off factor gets closer to zero, the transition becomes steeper, and the filter approaches the ideal low-pass filter.

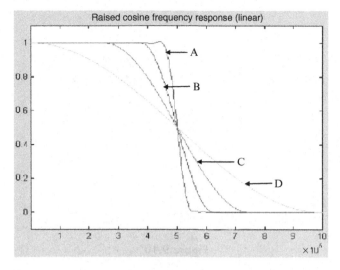

Figure 9.8

The impulse response of the raised cosine filter is shown in the following figures. Figure 9.9 shows the filter impulse response, and Figure 9.10 zooms in to better show the lobes of the filter impulse. Again, this response is plotted for several different roll-off factors. It is similar to the sinc impulse response in that it has zero crossings at time intervals of T (because this is shown in the sample domain, rather than the time domain, this similarity is not readily apparent from the figures).

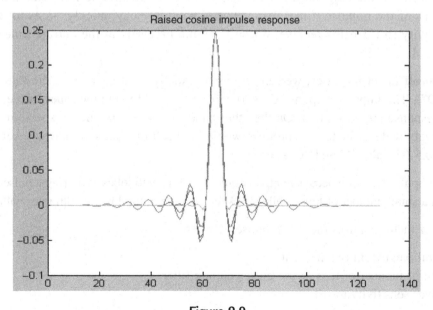

Figure 9.9

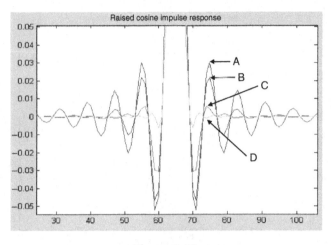

Figure 9.10

As the excess bandwidth is reduced to approach the ideal low-pass filter frequency response, the lobes in the impulse response become higher, approaching the sinc impulse response. The signal with the smaller amplitude lobes has a larger excess bandwidth, or wider spectrum.

Let's review this idea of pulse-shaping filter again, in light of Figures 9.9 and 9.10. We need a pulse-shaping filter that has a zero response at intervals of T in time so that a given symbol's pulse response does not have a contribution to the signal at the sampling times of the neighboring symbols. We also would like to minimize the height of the lobes of the impulse (time) response and have it decay quickly, so as to reduce our sensitivity to ISI if the receiver doesn't sample precisely at the correct time for each symbol.

As the roll off factor increases, we can see this is exactly what happens in the figures (signal "D"). The impulse response goes to zero very quickly, and the lobes of the filter impulse response are very small. On the other hand, we have a frequency spectrum which is excessively wide. A better compromise would be a roll off factor somewhere between 0.25 and 0.5 (signals "B" and "C").

Here, the impulse response decays relatively quickly with small lobes, requiring a pulse-shaping filter with a small number of taps, while still keeping the required bandwidth reasonable.

The roll-off factor controls the compromise between

- Spectral bandwidth requirement
- Length or number of taps of the pulse-shaping filter
- Receiver sensitivity to ISI

Another significant aspect of the transmit pulse-shaping filter is that it is always an interpolating filter. In our figures, this is shown as a 4× interpolation filter. If you look carefully at the impulse response in Figure 9.10, you can see that the zero crossings occur every four samples. This corresponds to $t = N \cdot T$ in the time domain, due to the 4× interpolation.

The transmit pulse-shaping filter must be an interpolating filter because our I and Q baseband signals must meet the Nyquist criterion. In this example, the symbol rate is 1 MSPS. If we use a high roll-off factor, the baseband spectrum of the I and Q signals can be as high as 1 MHz. So we require a minimum sampling rate of 2 MHz, or twice the symbol rate. Therefore, the pulse-shaping filter needs to interpolate by at least a factor of two, and is often interpolated quite a bit higher than this, for reasons we discuss in Chapter 11 on digital upconversion.

Once we have our pulse-shaped and interpolated I and Q baseband digital signals, we can use digital-to-analog converters (DACs) to create the analog I and Q baseband signals. These signals can be used to drive an analog mixer, which can create a passband signal. A passband signal is a baseband signal that has been upconverted or mixed with a carrier frequency.

For example, we might use a 0.25 roll-off filter for our 1 MSPS modulator. The baseband I and Q signals then have a bandwidth of 625 kHz. If we use a carrier frequency of 1 GHz, our transmit signal then requires about 1.25 MHz of spectrum centered at 1 GHz.

So far, we have discussed the process that occurs in the transmission path. The receive path is quite similar. The signal is downconverted, or mixed down to baseband. We discuss this topic in more detail in Chapter 11. The demodulation process starts with baseband I and Q signals. The receiver is more complex because it must deal with several additional issues. For example, there may be nearby signals that can interfere with the demodulation process. These signals must be filtered out, usually with a combination of analog and digital filters. The final stage of digital filtering is often the same pulse-shaping filter used in the transmitter. This is called a matched filter. The idea is that if the same filter that was used to create the signal is also used to filter the spectrum prior to sampling, we can maximize the amount of signal energy used in the detection (or sampling process). There is a bit of mathematics to prove this point, so we can just take it at face value. Due to this idea of using the same filter in the transmitter and receiver, the raised cosine filter is usually modified to a square root raised cosine filter. The frequency response of the raised cosine filter is modified to be the square root of the amplitude across the passband. This also modifies the impulse response as well. This is shown in Figures 9.11 and 9.12 for the same roll-off factors.

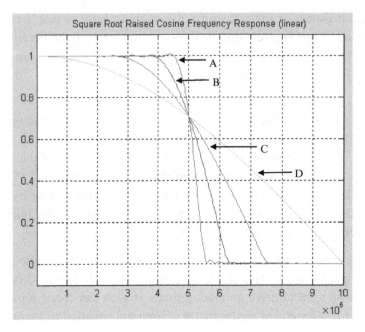

Figure 9.11

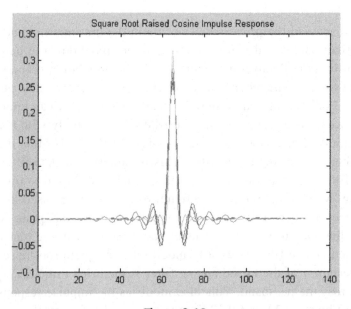

Figure 9.12

Since the signal passes through both filters, the net frequency response is the raised cosine filter. After passing through the receive pulse-shaping (also called matched) filter, the signal is sampled. Using the sampled I and Q value, the receiver chooses the constellation point in the I-Q plane closest to the sampled value. The bits corresponding to that symbol are recovered, and if all goes well, the receiver chooses the same symbol point selected by the transmitter. We have greatly simplified this whole process, but this is the essence of digital communications.

We can see why it will be easier to have errors when transmitting 64-QAM as compared to QPSK. The receiver has 64 closely spaced symbols to select from in the case of 64-QAM, whereas in QPSK, there are only 4 widely spaced symbols to select from. This makes 64-QAM systems much more susceptible to ISI, noise, or interference. You might think you should transmit the 64-QAM signal with higher power, to spread the symbols further apart. This is an effective but very expensive way to mitigate the noise and interference that prevents correct detection of the symbol at the receiver. Also, the transmit power is often limited by the regulatory agencies, or the transmitter may be battery powered or have other constraints.

The receiver also has a number of other problems to contend with. We assumed that we always sample at the correct instant in time when one symbol has a nonzero value in the signal. The receiver must somehow determine this correct sampling time, usually by a combination of trial and error during the initial part of the reception, and sometimes by having the transmitter send a predetermined (or training) sequence known by both transmitter and receiver. This process is known as acquisition, where the receiver tries to fine-tune the sampling time, the symbol rate, the exact frequency and phase of the carrier, and other parameters that may be needed to demodulate the received signal with a minimum of errors. And once all this information is determined, it must still be tracked to account for differences in transmit and receive clocks, Doppler shifts due to relative motion between the receiver and transmitter, and changes in the path the signal takes from transmitter to receiver, causing various reflections, distortions, and attenuations.

These problems are what make digital receivers so difficult and interesting to work with. Unfortunately, there is usually a lot of mathematics associated with most receiver algorithms and methods, so we do not go into this topic in any depth. But later chapters describe the basic principles of several common types of digital communication systems.

Figures 9.13 and 9.14 show plots from both a 16-QAM and 64-QAM constellation after being sampled by an actual digital receiver. Each receiver signal has the same average energy. This is from a WiMax wireless system, operating in the presence of noise. The receiver does manage to do a sufficiently good job at detection so that each constellation point is clearly visible. But we can imagine that as the receiver noise level increases, the constellation samples would quickly start to drift together on the 64-QAM constellation, and we would be unable to accurately determine to which constellation point a given symbol should map. Error vector magnitude (EVM) is a measurement of the constellation noise level. The 16-QAM system is more robust in the presence of additive noise and other impairments, compared to the 64-QAM.

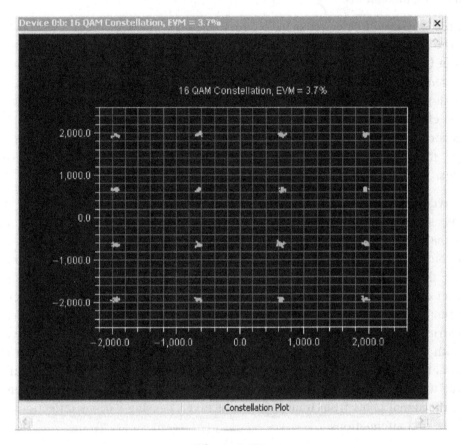

Figure 9.13

The modulation and demodulation (modem) ideas presented in this chapter are used in most digital communication systems, including satellite, microwave, cellular (CDMA and TDMA), wireless LAN (OFDM), DSL, fax, and data dial-up modems. Actually, the lowly dial-up modem is among the most complicated of all: a V.34 modem can have over 1000 constellation points.

I hope that by now the name conventions of the modulation methods are starting to make more sense to you. In QPSK, all four symbols have the same amplitude. The phase in the

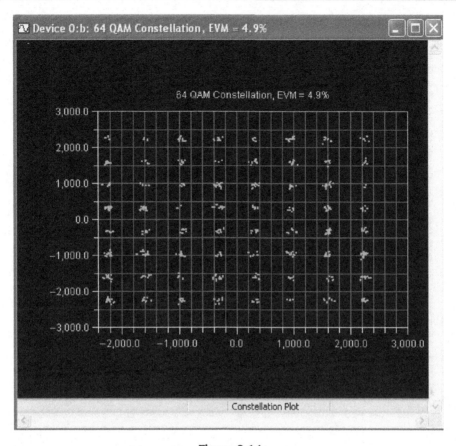

Figure 9.14

complex plane is what distinguishes the different symbols, each of which is located in a different quadrant. For QAM, the amplitude and phase of the symbol are needed to distinguish a particular symbol.

In general, communication systems are full of trade-offs. The most important comes from a famous theorem developed by Claude Shannon that gives the maximum theoretical data bit rate which can be communicated over a communications channel depending on bandwidth, transmit power, and receiver noise level. It gives the maximum data rate that can be sent over a channel or link with a given noise level and bandwidth. This is known as the Shannon limit, and is somewhat analogous to the speed of light, which can be approached with ever-increasing amounts of cleverness and effort but can never be exceeded. This topic is discussed further in Chapter 12 on error correction codes.

Figure 5.14

CHAPTER 10

Discrete and Fast Fourier Transforms (DFT, FFT)

You have likely heard the term *FFT*. In this chapter, we discuss the discrete Fourier transform, or DFT, and its more popular cousin, the fast Fourier transform, or FFT. We again try to approach this topic intuitively, although a bit of math is unavoidable.

The key is to understand the DFT operation. The FFT is simply a highly optimized implementation of the DFT. They both produce identical results. While the implementation of the FFT is interesting, unless you are actually building it, you don't really need to know the details. Because the FFT is such a basic building block in DSP, many implementations are already available.

So just what is the DFT? To start with, the DFT is simply a transform. It takes a sequence of sampled data (a signal) and computes the frequency content of that sampled data sequence. This gives the representation of the signal in the frequency domain, as opposed to the familiar time domain representation. This process is very similar to what we did when computing frequency response of FIR filters. This tool can be very powerful in signal processing applications because it allows us to examine any given signal (not just a filter) in the frequency domain, which provides the spectral content of a given signal.

There is also an inverse discrete Fourier transform (IDFT) and inverse fast Fourier transform (IFFT). Again, the IFFT is simply an optimized form of the IDFT. They both compute the time domain representation of the signal from the frequency domain information. Using these transforms, we can go back and forth from the time domain signal and the frequency domain spectral representation.

Conceptually, the DFT tries to take any complex signal, and break it up into a sum of many cosine and sine waves of different frequencies.

This idea is a bit novel if this is your first exposure to it.

To appreciate what is happening, we are going to examine a few simple examples. Doing so involves multiplying and summing up complex numbers, which, while not difficult, can be tedious. We minimize the tedium by using a short transform length, but it cannot

Digital Signal Processing 101. DOI: 10.1016/B978-1-85617-921-8.00014-6

really be avoided if you want to understand the DFT and later if you wish to delve into FFT (the optimized form of DFT).

Let's begin with the equation we used earlier for frequency response in Chapter 5 on FIR filters:

$$\text{Frequency response} = H(\omega) = \sum_{i=-\infty \text{ to } \infty} C_i\, e^{-j\omega i}$$

Let's see if we can start simplifying. For example, say we decide to perform the calculation over a finite length of sampled data signal "x," which contains N samples rather than an infinite sequence of coefficients C_i. Doing so gets rid of the infinity and makes this something we can actually build:

$$H(\omega) = \sum_{i=0 \text{ to } N-1} X_i e^{-j\omega i}$$

Next, notice that ω is a continuous variable, which we evaluate over a 2π interval, usually from $-\pi$ to π. Instead, we will transform a sampled time domain signal to a sampled frequency domain spectral plot. So rather than computing ω continuously from $-\pi$ to π, we instead compute ω at M equally spaced points over an interval of 2π. Now it turns out that to avoid aliasing in the frequency domain, we must make $M \geq N$.

The reverse transform is the IDFT (or IFFT), which reconstructs the sampled time domain signal of length N. This process requires N or more points in the frequency domain.

The end result is that we set $N = M$, and the frequency domain representation has the same number of points as the time domain representation. We use the convention "x_i" to represent the time domain version of the signal, and "X_k" to represent the frequency domain representation of the signal. Both indexes "i" and "k" run from 0 to $N - 1$.

10.1 DFT and IDFT Equations

The DFT and IDFT equations appear very similar:

$$\text{DFT(time} \rightarrow \text{frequency)} \quad X_k = II(2\pi k/N) - \sum_{i=0 \text{ to } N-1} X_i e^{-j2\pi ki/N}, \text{ for } k = \{0, \ldots, N-1\}$$

$$\text{DFT(frequency} \rightarrow \text{time)} \quad X_i = 1/N \cdot \sum_{k=0 \text{ to } N-1} X_k e^{+j2\pi ki/N}, \text{ for } i = \{0, \ldots, N-1\}$$

The differences are the negative sign on the exponent on the DFT equation and the factor of 1/N on the IDFT equation. The DFT equation requires that every single sample in the frequency domain has a contribution from each and every one of the time domain samples. And the IDFT equation requires that every single sample in the time domain has a contribution from each and every one of the frequency domain samples. To compute a single sample of either transform requires N complex multiplication and addition operations.

To compute the entire transform requires computing N samples, for a total of N^2 multiplication and addition operations. This can become a computational problem when N grows large. As we see later, this is the reason the FFT and IFFT were developed.

The values of X_k represent the amount of signal energy at each frequency point. Imagine taking a spectrum of 1 MHz. Then we divide it into N bins. If $N = 20$, then we will have 20 frequency bins, each 50 kHz wide. The DFT output, X_k, represents the signal energy in each of these bins. For example, X_0 represents the signal energy at 0 kHz, or the DC component of the signal. X_1 represents the frequency content of the signal at 50 kHz. X_2 represents the frequency content of the signal at 100 kHz. X_{19} represents the frequency content of the signal at 950 kHz.

Now let's examine a few comments on these transforms. First, they are reversible. We can take a signal represented by N samples and perform the DFT on it. We get N outputs representing the frequency response or spectrum on the signal. If we take this frequency response and perform the IDFT on it, we get back our original signal of N samples. Second, when the DFT output gives the frequency content of the input signal, it assumes that the input signal is periodic in N. To put it another way, the frequency response is actually the frequency response of an infinite long periodic signal, where the N long sequence of x_i samples repeats over and over. Last, the input signal x_i is usually assumed to be a complex (two-dimensional) signal. The frequency response samples X_k are also complex. Often we are more interested in only the magnitude of the frequency response X_k, which can be more easily displayed. But to get back the original complex input x_i using the IDFT, we would need the complex sequence X_k.

At this point, let's examine a few examples, selecting $N = 8$.

For our $N = 8$-point DFT, the output gives us the distribution of input signal energy into 8 frequency bins, corresponding to the frequencies in the following table. By computing the DFT coefficients X_k, we are performing a correlation, or trying to match our input signal to each of these frequencies. The magnitude DFT output coefficients X_k represent the degree of match of the time domain signal x_i to each frequency component.

Table 10.1: 8-Point DFT coefficients

k	X_k	Compute by Correlating to Complex Exponential Signal	ΔPhase Between Each Sample of Complex Exponential Signal
0	X_0	e^0 for $i = 0, 1, \ldots, 7$	0
1	X_1	$e^{-j2\pi i/8}$ for $i = 0, 1, \ldots, 7$	$-\pi/4$ or -45 degrees
2	X_2	$e^{-j4\pi i/8}$ for $i = 0, 1, \ldots, 7$	$-2\pi/4$ or -90 degrees
3	X_3	$e^{-j6\pi i/8}$ for $i = 0, 1, \ldots, 7$	$-3\pi/4$ or -135 degrees
4	X_4	$e^{-j8\pi i/8}$ for $i = 0, 1, \ldots, 7$	$-4\pi/4$ or -180 degrees
5	X_5	$e^{-j10\pi i/8}$ for $i = 0, 1, \ldots, 7$	$-5\pi/4$ or -225 degrees
6	X_6	$e^{-j12\pi i/8}$ for $i = 0, 1, \ldots, 7$	$-6\pi/4$ or -270 degrees
7	X_7	$e^{-j14\pi i/8}$ for $i = 0, 1, \ldots, 7$	$-7\pi/4$ or -315 degrees

10.1.1 First DFT Example

Let us start with a simple time domain signal consisting of $\{1,1,1,1,1,1,1,1\}$. Remember, the DFT assumes this signal keeps repeating, so the frequency output will actually be that of an indefinite string of 1s. Because this signal is unchanging, then by intuition we expect that the zero frequency component (DC of signal) is going to be the only nonzero component of the DFT output X_k.

Starting with $X_k = \displaystyle\sum_{i=0 \text{ to } N-1} x_i\, e^{-j2\pi ki/N}$ and setting $N = 8$ and all $x_i = 1$

$$X_k = \sum_{i=0 \text{ to } 7} 1 \cdot e^{-j2\pi ki/8},$$

and setting $k = 0$ (recall that $e^0 = 1$)

$$X_0 = \sum_{i=0 \text{ to } 7} 1 \cdot 1 = 8$$

Next, we evaluate for $k = 1$:

$$\begin{aligned}
X_1 &= \sum_{i=0 \text{ to } 7} 1 \cdot e^{-j2\pi i/8} \\
&= 1 + e^{-j2\pi/8} + e^{-j4\pi/8} + e^{-j6\pi/8} + e^{-j8\pi/8} + e^{-j10\pi/8} + e^{-j12\pi/8} + e^{-j14\pi/8}
\end{aligned}$$

$$\begin{aligned}
X_1 = &\ 1 + (0.7071 - j0.7071) - j + (-0.7071 - j0.7071) - 1 + (-0.7071 - j0.7071) \\
&+ j + (0.7071 - j0.7071)
\end{aligned}$$

$$X_1 = 0$$

The eight terms of the summation for X_1 cancel out. This makes sense if you think about it. This is a sum of eight equally spaced points about the origin on the unit circle of complex plane. The summation of these points must equal the center—in this case, zero.

Next, we evaluate for $k = 2$:

$$X_2 = \sum_{i=0 \text{ to } 7} 1 \cdot e^{-j2\pi i/8} = 1 + e^{-j\pi/2} + e^{-j\pi} + e^{-j3\pi/2} + e^{-j2\pi} + e^{-j5\pi/2} + e^{-j3\pi} + e^{-j7\pi/2}$$

$$X_2 = 1 - j - 1 + j + 1 - j - 1 + j = 0$$

We find out similarly that X_3, X_4, X_5, X_6, and X_7 also are zero. Each of these represents eight points equally spaced about the unit circle. X_1 has its points spaced at $-45°$ increments, X_2 has its points spaced at $-90°$ increments, X_3 has its points spaced at $-135°$ increments, and so forth (the points may wrap around the unit circle in the frequency domain multiple times). So as we expected, the only nonzero term is X_0, which is the DC term. There is no other frequency content of the signal.

Now, let us use the IDFT to get back the original sequence:

$$x_i = 1/N \cdot \sum_{k=0 \text{ to } N-1} X_k e^{+j2\pi ki/N} \quad \text{for } N = 8 \text{ and } X_0 = 8, \text{ all other } X_k = 0$$

$$x_i = 1/8 \cdot \sum_{k=0 \text{ to } N-1} X_k e^{+j2\pi ki/8}$$

Since $X_0 = 8$ and the rest are zero, we need to evaluate only the summation for $k = 0$:

$$x_i = 1/8 \cdot 8 \cdot e^{+j2\pi 0 i/8} = 1$$

This is true for all values of i (the 0 in the exponent means the value of i is irrelevant). So we get an infinite sequence of 1s.

In general, however, we would evaluate for i from 0 to N − 1. Due to the periodicity of the transform, there is no point in evaluating when i = N or greater. If we evaluate for i = N, we get the same value as i = 0, and for i = N + 1, we get the same value as i = 1.

10.1.2 Second DFT Example

Let us consider another simple example, with a time domain signal $\{1, j, -1, -j, 1, j, -1, -j\}$. This is actually the complex exponential $e^{+j2\pi i/4}$. This signal consists of a single frequency and corresponds to one of the frequency "bins" that the DFT will measure. So we can expect a nonzero DFT output in this frequency bin, but zero elsewhere. Let's see how this works out.

Starting with $X_k = \sum_{i=0 \text{ to } N-1} x_i e^{-j2\pi ki/N}$ and setting N = 8 and $x_i = \{1, j, -1, -j, 1, j, -1, -j\}$

$$X_0 = \sum_{i=0 \text{ to } 7} x_i \cdot 1, \text{ as } k = 0 (e^0 = 1)$$

$X_0 = 1 + j - 1 - j + 1 + j - 1 - j = 0$, so the signal has no DC content, as expected. Notice that to calculate X_0, which is the DC content of x_i, the DFT reduces to just summing (essentially averaging) the input samples.

Next, we evaluate for k = 1:

$$X_1 = \sum_{i=0 \text{ to } 7} x_i \cdot e^{-j2\pi i/8} \quad = 1 \cdot 1 + j \cdot e^{-j2\pi/8} - 1 \cdot e^{-j4\pi/8} - j \cdot e^{-j6\pi/8}$$

$$+ 1 \cdot e^{-j8\pi/8} + j \cdot e^{-j10\pi/8} - 1 \cdot e^{-j12\pi/8} - j \cdot e^{-j14\pi/8}$$

$$X_1 = 1 + [j \cdot (0.7071 - j0.7071)] + j - [j \cdot (-0.7071 - j0.7071)]$$
$$-1 + [j \cdot (-0.7071 + j0.7071)] - j - [j \cdot (0.7071 + j0.7071)]$$

$$X_1 = 0$$

If you take the time to work this out, you see that all eight terms of the summation cancel out. This also happens for X_3, X_4, X_5, X_6, and X_7. Let's look at X_2 now. We also express x_i using the complex exponential format of $e^{+j2\pi i/4}$:

$$X_k = \sum_{i=0 \text{ to } 7} x_i e^{-j2\pi ki/8}$$

$$X_2 = \sum_{i=0 \text{ to } 7} x_i \cdot e^{-j4\pi i/8} = \sum_{i=0 \text{ to } 7} e^{+j2\pi i/4} \cdot e^{-j4\pi i/8} = \sum_{i=0 \text{ to } 7} e^{+j2\pi i/4} \cdot e^{-j2\pi i/4}$$

Remember that when exponentials are multiplied, the exponents are added ($x^2 \cdot x^3 = x^5$). Here, the exponents are identical, except of opposite sign. So they add to zero:

$$X_2 = \sum_{i=0 \text{ to } 7} e^{+j2\pi i/4} \cdot e^{-j2\pi i/4} \sum_{i=0 \text{ to } 7} e^0 = \sum_{i=0 \text{ to } 7} 1 = 8$$

The sole frequency component of the input signal is X_2. The reason is that our input is a complex exponential frequency at the exact frequency that X_2 represents.

10.1.3 Third DFT Example

Next, we can try modifying x_i such that we introduce a phase shift or delay (like substituting a sine wave for a cosine wave). Suppose we introduce a delay, so x_i starts at j instead of 1, but is still the same frequency. The input x_i is still rotating around the complex plane at the same rate but starts at j (angle of $\pi/2$) rather than 1 (angle of 0). Now the sequence $x_i = \{j, -1, -j, 1, j, -1, -j, 1\}$ or $e^{+j(2\pi(i+1)/4)}$.

The DFT output results in X_0, X_1, X_3, X_4, X_5, X_6, and $X_7 = 0$, as before. Changing the phase cannot cause any new frequency to appear in the other bins.

Next, we evaluate for k = 2:

$$X_k = \sum_{i=0 \text{ to } 7} x_i e^{-j2\pi ki/8}$$

$$X_2 = \sum_{i=0 \text{ to } 7} x_i \cdot e^{-j4\pi i/8} = \sum_{i=0 \text{ to } 7} e^{+j(2\pi(i+1)/4)+1)} \cdot e^{-j4\pi i/8}$$

We need to sum the two values of the two exponents:

$$+j(2\pi(i+1)/4) + -j4\pi i/8 = +j2\pi i/4 + j2\pi/4 - j2\pi i/4 = j\pi/2$$

Substituting back this exponent value

$$X_2 = \sum_{i=0 \text{ to } 7} e^{+j((2\pi i/4)+1)} \cdot e^{-j4\pi i/8} = \sum_{i=0 \text{ to } 7} e^{+j\pi/2} = \sum_{i=0 \text{ to } 7} j = j8$$

So we get exactly the same magnitude at the frequency component X_2. The difference is the phase of X_2. So the DFT does not just pick out the frequency components of a signal but is sensitive to the phase of those components. The phase as well as amplitude of the frequency components X_k can be represented because the DFT output is complex.

The process of the DFT is to correlate the N sample input data stream x_i against N equally spaced complex frequencies. If the input data stream is one of these N complex frequencies, then we get a perfect match and get zero in the other N − 1 frequencies that do not match. But what happens if we have an input data stream with a frequency in between one of the N frequencies?

To review, we have looked at three simple examples. The first was a constant-level signal, so the DFT output was just the zero frequency or DC component. The second example was a complex frequency that matched exactly to one of the frequency bins, X_k, of the DFT. The third was the same complex frequency, but with a phase offset. The fourth, examined in the following section, is a complex frequency not matched to one of the N frequencies used by the DFT. Next, we look at an example where the frequency is somewhere in between the DFT bins.

10.1.4 Fourth DFT Example

Now let's look at an input signal of frequency $e^{+j2.1\pi i/8}$. This is pretty close to $e^{+j2\pi i/8}$, so we would expect a pretty strong output at X_1. Let's see what the N = 8 DFT result is; let's hope the arithmetic is all correct. Slogging through this arithmetic is purely optional; the details are shown to provide a complete example.

Generic DFT equation for N = 8: $X_k = \sum_{i=0 \text{ to } 7} x_i e^{-j2\pi ki/8}$

$$X_0 = \sum_{i=0 \text{ to } 7} e^{+j2.1\pi i/8} \cdot 1 = \sum_{i=0 \text{ to } 7} e^{+j2.1\pi i/8}$$

$$= [1 + j0] + [0.6788 + j0.7343] + [-0.0785 + j0.9969] + [-0.7853 + j0.6191]$$

$$+ [-0.9877 - j0.1564] + [-0.5556 - j0.8315] + [0.2334 - j0.9724] + [0.8725 - j0.4886]$$

$$= 0.3777 - j0.0986$$

$$X_1 = \sum_{i=0 \text{ to } 7} e^{+j2.1\pi i/8} \cdot e^{-j2\pi i/8} = \sum_{i=0 \text{ to } 7} e^{+j0.1\pi i/8}$$

$$= [1 + j0] + [0.9992 + j0.0393] + [0.9969 + j0.0785] + [0.9931 + j0.1175]$$

$$+ [0.9877 + j0.1564] + [0.9808 + j0.1951] + [0.9724 + j0.2334] + [0.9625 + j0.2714]$$

$$= 7.8925 + j1.0917$$

$$X_2 = \sum_{i=0 \text{ to } 7} e^{+j2.1\pi i/8} \cdot e^{-j4\pi i/8} = \sum_{i=0 \text{ to } 7} e^{-j1.9\pi i/8}$$

$$= [1 + j0] + [0.7343 - j0.6788] + [0.0785 - j0.9969] + [-0.6191 - j0.7853]$$

$$+ [-0.9877 - j0.1564] + [-0.8315 + j0.5556] + [-0.2334 + j0.9724] + [0.4886 + j0.8725]$$

$$= -0.3703 - j0.2170$$

$$X_3 = \sum_{i=0 \text{ to } 7} e^{+j2.1\pi i/8} \cdot e^{-j6\pi i/8} = \sum_{i=0 \text{ to } 7} e^{-j3.9\pi i/8}$$

$$= [1 + j0] + [0.0393 - j0.9992] + [-0.9969 - j0.0785] + [-0.1175 + j0.9931]$$

$$+ [0.9877 + j0.1564] + [0.1951 - j0.9808] + [-0.9724 - j0.2334] + [-0.2714 + j0.9625]$$

$$= -0.1362 - j0.1800$$

$$X_4 = \sum_{i=0 \text{ to } 7} e^{+j2.1\pi i/8} \cdot e^{-j8\pi i/8} = \sum_{i=0 \text{ to } 7} e^{-j5.9\pi i/8}$$

$$= [1 + j0] + [-0.6788 - j0.7343] + [-0.0785 + j0.9969] + [0.7853 - j0.6191]$$

$$+ [-0.9877 - j0.1564] + [-0.5556 + j0.8315] + [0.2334 - j0.9724] + [-0.8725 + j0.4886]$$

$$= -0.0431 - j0.1652$$

$$X_5 = \sum_{i=0 \text{ to } 7} e^{+j2.1\pi i/8} \cdot e^{-j10\pi i/8} = \sum_{i=0 \text{ to } 7} e^{-j7.9\pi i/8}$$

$$= [1 + j0] + [-0.9992 - j0.0393] + [0.9969 + j0.0785] + [-0.9931 - j0.1175]$$

$$+ [0.9877 + j0.1564] + [-0.9808 - j0.1951] + [0.9724 + j0.2334] + [-0.9625 - j0.2714]$$

$$= 0.0214 - j0.1550$$

$$X_6 = \sum_{i=0 \text{ to } 7} e^{+j2.1\pi i/8} \cdot e^{-j12\pi i/8} = \sum_{i=0 \text{ to } 7} e^{-j9.9\pi i/8}$$

$$= [1 + j0] + [-0.7343 + j0.6788] + [0.0785 - j0.9969] + [0.6191 + j0.7853]$$

$$+ [-0.9877 - j0.1564] + [0.8315 - j0.5556] + [-0.2334 + j0.9724] + [-0.4886 - j0.8725]$$

$$= 0.0849 - j0.1449$$

$$X_7 = \sum_{i=0 \text{ to } 7} e^{+j2.1\pi i/8} \cdot e^{-j14\pi i/8} = \sum_{i=0 \text{ to } 7} e^{-j11.9\pi i/8}$$

$$= [1 + j0] + [-0.0393 + j0.9992] + [-0.9969 - j0.0785] + [0.1175 - j0.9931]$$

$$+ [0.9877 + j0.1564] + [-0.1951 + j0.9808] + [-0.9724 - j0.2334] + [0.2714 - j0.9625]$$

$$= 0.1730 - j0.1310$$

This is a bit tedious. But there is some insight to be gained from the results of these simple examples, as you can see in the following table.

This table shows how the DFT is able to represent the signal energy in each frequency bin. The first example has all its energy at DC. The second and third examples are complex exponentials at frequency $\omega = \pi/2$ rad/sample, which corresponds to DFT output X_2. The magnitude of the DFT outputs is the same for both examples, since the only difference of

Table 10.2: Example DFT results

DFT Output Magnitude	$x_i = \{1,1,1,1,1,1,1,1\}$	$x_i = e^{+j2\pi i/4}$	$x_i = e^{+j(2\pi(i+1)/4)}$	$x_i = e^{+j2.1\pi i/8}$
Output X_0	8	0	0	0.39
Output X_1	0	0	0	7.99
Output X_2	0	8	8	0.43
Output X_3	0	0	0	0.23
Output X_4	0	0	0	0.17
Output X_5	0	0	0	0.16
Output X_6	0	0	0	0.17
Output X_7	0	0	0	0.22

the inputs is the phase. The fourth example is the most interesting. In this case, the input frequency is close to $\pi/4$ rad/sample, which corresponds to DFT output X_1. So X_1 does capture most of the energy of the signal. But small amounts of energy spill into other frequency bins, particularly the adjacent bins.

We can increase the frequency sorting ability of the DFT by increasing the value of N. Then each frequency bin is narrower (since the frequency spectrum is divided in N sections in the DFT). This results in any given frequency component being more selectively represented by a particular frequency bin. For example, the frequency response plots of the filters contained in Chapter 5 on FIR filters are computed with a value of N equal to 1024. This means the spectrum was divided into 1024 sections, and the response computed for each particular frequency. When plotted together, this gives a very good representation of the complete frequency spectrum.

Note this also requires taking a longer input sample stream x_i, equal to N. This, in turn, requires a much greater number of operations to compute.

At some point, some smart people searched for a way to compute the DFT in a more efficient way. The result is the FFT, or fast Fourier transform. Rather than requiring N^2 complex multiplies and additions, the FFT requires $N \cdot \log_2 N$ complex multiplication and addition operations. This may not sound like a big deal, but look at the comparison in the following table.

Table 10.3: FFT computational efficiency

N	DFT – N^2 Complex Multiplication and Addition Operations	FFT – $N \cdot \log_2 N$ Complex Multiplication and Addition Operations	Computational Effort of FFT Compared to DFT (%)
8	64	24	37.50
32	1024	160	15.62
256	65,536	2048	3.12
1024	1,048,576	10,240	0.98
4096	16,777,216	49,152	0.29

So by using the FFT algorithm on a 1024-point (or sample) input, we are able to reduce the computational requirements to less than 1%, or by 2 orders of magnitude, of what the DFT algorithm would require.

10.2 Fast Fourier Transform (FFT)

Let's start with the calculation of the simplest DFT: $N = 2$ DFT.

Generic DFT equation for $N = 2$: $X_k = \sum_{i=0 \text{ to } 1} x_i e^{-j2\pi ki/2}$

$$X_0 = x_0 \cdot e^{-j2\pi 0/2} + x_1 \cdot e^{-j2\pi 0/2}$$
$$X_1 = x_0 \cdot e^{-j2\pi 0/2} + x_1 \cdot e^{-j2\pi 1/2}$$

Simplifying since $e^0 = 1$, we find

$$X_0 = x_0 + x_1$$
$$X_1 = x_0 + x_1 \cdot e^{-j\pi} = x_0 - x_1$$

Next, we do the 4-point ($N = 4$) DFT.

Generic DFT equation for $N = 4$: $X_k = \sum_{i=0 \text{ to } 3} x_i e^{-j2\pi ki/4}$

$$X_0 = x_0 \cdot e^{-j2\pi 0/4} + x_1 \cdot e^{-j2\pi 0/4} + x_2 \cdot e^{-j2\pi 0/4} + x_3 \cdot e^{-j2\pi 0/4}$$

$$X_1 = x_0 \cdot e^{-j2\pi 0/4} + x_1 \cdot e^{-j2\pi 1/4} + x_2 \cdot e^{-j2\pi 2/4} + x_3 \cdot e^{-j2\pi 3/4}$$

$$X_2 = x_0 \cdot e^{-j2\pi 0/4} + x_1 \cdot e^{-j2\pi 2/4} + x_2 \cdot e^{-j2\pi 4/4} + x_3 \cdot e^{-j2\pi 6/4}$$

$$X_3 = x_0 \cdot e^{-j2\pi 0/4} + x_1 \cdot e^{-j2\pi 3/4} + x_2 \cdot e^{-j2\pi 6/4} + x_3 \cdot e^{-j2\pi 9/4}$$

The term $e^{-j2\pi k/4}$ repeats itself with a period of $k = 4$ because the complex exponential makes a complete circle and begins another. This periodicity means that $e^{-j2\pi k/4}$ is equal when evaluated for $k = 0, 4, 8, 12, \ldots$. It is again equal for $k = 1, 5, 9, 13, \ldots$. Consequently, we can simplify the last two terms of expressions for X_2 and X_3 (shown in bold below). We can also remove the exponential when it is to a power of zero.

$$X_0 = x_0 + x_1 + x_2 + x_3$$

$$X_1 = x_0 + x_1 \cdot e^{-j2\pi 1/4} + x_2 \cdot e^{-j2\pi 2/4} + x_3 \cdot e^{-j2\pi 2/4}$$

$$X_2 = x_0 + x_1 \cdot e^{-j2\pi 2/4} + x_2 \cdot e^{-j2\pi 0/4} + x_3 \cdot e^{-j2\pi 2/4}$$

$$X_3 = x_0 + x_1 \cdot e^{-j2\pi 3/4} + x_2 \cdot e^{-j2\pi 2/4} + x_3 \cdot e^{-j2\pi 1/4}$$

Now we are going to rearrange the terms of the 4-point ($N = 4$) DFT. The even and odd terms are grouped together:

$$X_0 = [x_0 + x_2] + [x_1 + x_3]$$

$$X_1 = [x_0 + x_2^{-j2\pi2/4}] + [x_1 \cdot e^{-j2\pi1/4} + x_3 \cdot e^{-j2\pi3/4}]$$

$$X_2 = [x_0 + x_2] + [x_1 \cdot e^{-j2\pi2/4} + x_3 \cdot e^{-j2\pi2/4}]$$

$$X_3 = [x_0 + x_2 \cdot e^{-j2\pi2/4}] + [x_1 \cdot e^{-j2\pi3/4} + x_3 \cdot e^{-j2\pi1/4}]$$

Next, we factor x_1 and x_3 to get this particular form:

$$
\begin{aligned}
X_0 &= [x_0 + x_2] + [x_1 + x_3]\\
X_1 &= [x_0 + x_2 \cdot e^{-j2\pi2/4}] + [x_1 \cdot e^{-j2\pi1/4} + x_3 \cdot e^{-j2\pi3/4}]\\
&= [x_0 + x_2 \cdot e^{-j2\pi2/4}] + [x_1 + x_3 \cdot e^{-j2\pi2/4}] \cdot e^{-j2\pi/4}\\
X_2 &= [x_0 + x_2] + [x_1 \cdot e^{-j2\pi2/4} + x_3 \cdot e^{-j2\pi2/4}]\\
&= [x_0 + x_2] + [x_1 + x_3] \cdot e^{-j2\pi2/4}\\
X_3 &= [x_0 + x_2 \cdot e^{-j2\pi2/4}] + [x_1 \cdot e^{-j2\pi3/4} + x_3 \cdot e^{-j2\pi1/4}]\\
&= [x_0 + x_2 \cdot e^{-j2\pi2/4}] + [x_1 + x_3 \cdot e^{-j2\pi2/4}] \cdot e^{-j2\pi3/4}
\end{aligned}
$$

Here is the result:

$$
\begin{aligned}
\mathbf{X_0} &= \mathbf{[x_0 + x_2] + [x_1 + x_3]}\\
X_1 &= [x_0 + x_2 \cdot e^{-j2\pi2/4}] + [x_1 + x_3 \cdot e^{-j2\pi2/4}] \cdot e^{-j2\pi/4}\\
X_2 &= [x_0 + x_2] + [x_1 + x_3] \cdot e^{-j4\pi/4}\\
X_3 &= [x_0 + x_2 \cdot e^{-j2\pi2/4}] + [x_1 + x_3 \cdot e^{-j2\pi2/4}] \cdot e^{-j6\pi/4}
\end{aligned}
$$

Now comes the insightful part. Comparing the preceding four equations, you can see that the bracketed terms used for X_0 and X_1 are also present in X_2 and X_3. So we do not need to recompute these terms during the calculation of X_2 and X_3. We can simply multiply them by the additional exponential outside the brackets. This reusing of partial products in multiple calculations is the key to understanding the FFT efficiency, so at the risk of being repetitive, this example is shown again more explicitly here:

$$\rightarrow \quad \text{define } A = [x_0 + x_2], B = [x_1 + x_3]$$

$$X_0 = [x_0 + x_2] + [x_1 + x_3] = A + B$$

$$\rightarrow \quad \text{define } C = [x_0 + x_2 \cdot e^{-j2\pi2/4}], D = [x_1 + x_3 \cdot e^{-j2\pi2/4}]$$

$$X_1 = [x_0 + x_2 \cdot e^{j2\pi2/4}] + [x_1 + x_3 \cdot e^{-j2\pi2/4}] \cdot e^{-j2\pi/4} = C + D \cdot e^{-j2\pi/4}$$

$$X_2 = [x_0 + x_2] + [x_1 + x_3] \cdot e^{-j4\pi/4} = A + B \cdot e^{-j4\pi/4}$$

$$X_3 = [x_0 + x_2 \cdot e^{-j2\pi2/4}] + [x_1 + x_3 \cdot e^{-j2\pi2/4}] \cdot e^{-j6\pi/4} = C + D \cdot e^{-j6\pi/4}$$

This process quickly gets out of hand for anything larger than a 4-point (N = 4) FFT. So we are going to use a type of representation called a flow graph, as shown in Figure 10.1.

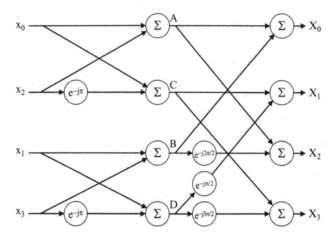

Figure 10.1

The flow graph is an equivalent way of representing the equations and, moreover, represents the actual organization of the computations. You should check for yourself in the preceding simple example that the flow graph gives the same results as the DFT equations. For example, $X_0 = x_0 + x_1 + x_2 + x_3$, and by examining the flow graph, you can see that $X_0 = A + B = [x_0 + x_2] + [x_1 + x_3]$, which is the same result. The order of computations would be to compute pairs {A,C} and {B,D} in the first stage. The next stage would be to compute {X_0,X_2} and {X_1,X_3}.

These stages (the preceding example has two stages) are composed of "butterflies." Each butterfly has two complex inputs and two complex outputs. The butterfly involves one or two complex multiplication and two complex addition operations. In the first stage, there are two butterflies to compute the two pairs {A,C} and {B,D}. In the second stage, there are two butterflies to compute the two pairs {X_0, X_2} and {X_1,X_3}. The complex exponentials multiplying the data path are known as "twiddle factors." In higher N count FFTs, they are simply sine and cosine values. These values are usually stored in a table.

Although you may grumble, next we are going to present an 8-point FFT:

$$\text{Generic FDT equation for N} = 8: \quad X_k = \sum_{i=0 \text{ to } 7} X_i e^{-j2\pi ki/8}$$

$$X_0 = x_0 \cdot e^{-j2\pi 0/8} + x_1 \cdot e^{-j2\pi 0/8} + x_2 \cdot e^{-j2\pi 0/8} + x_3 \cdot e^{-j2\pi 0/8}$$
$$+ x_4 \cdot e^{-j2\pi 0/8} + x_5 \cdot e^{-j2\pi 0/8} + x_6 \cdot e^{-j2\pi 0/8} + x_7 \cdot e^{-j2\pi 0/8}$$

$$X_1 = x_0 \cdot e^{-j2\pi 0/8} + x_1 \cdot e^{-j2\pi 1/8} + x_2 \cdot e^{-j2\pi 2/8} + x_3 \cdot e^{-j2\pi 3/8}$$
$$+ x_4 \cdot e^{-j2\pi 4/8} + x_5 \cdot e^{-j2\pi 5/8} + x_6 \cdot e^{-j2\pi 6/8} + x_7 \cdot e^{-j2\pi 7/8}$$

$$X_2 = x_0 \cdot e^{-j2\pi 0/8} + x_1 \cdot e^{-j2\pi 2/8} + x_2 \cdot e^{-j2\pi 4/8} + x_3 \cdot e^{-j2\pi 6/8}$$
$$+ x_4 \cdot e^{-j2\pi 8/8} + x_5 \cdot e^{-j2\pi 10/8} + x_6 \cdot e^{-j2\pi 12/8} + x_7 \cdot e^{-j2\pi 14/8}$$

$$X_3 = x_0 \cdot e^{-j2\pi 0/8} + x_1 \cdot e^{-j2\pi 3/8} + x_2 \cdot e^{-j2\pi 6/8} + x_3 \cdot e^{-j2\pi 9/8}$$
$$+ x_4 \cdot e^{-j2\pi 12/8} + x_5 \cdot e^{-j2\pi 15/8} + x_6 \cdot e^{-j2\pi 18/8} + x_7 \cdot e^{-j2\pi 21/8}$$

$$X_4 = x_0 \cdot e^{-j2\pi 0/8} + x_1 \cdot e^{-j2\pi 4/8} + x_2 \cdot e^{-j2\pi 8/8} + x_3 \cdot e^{-j2\pi 12/8}$$
$$+ x_4 \cdot e^{-j2\pi 16/8} + x_5 \cdot e^{-j2\pi 20/8} + x_6 \cdot e^{-j2\pi 24/8} + x_7 \cdot e^{-j2\pi 28/8}$$

$$X_5 = x_0 \cdot e^{-j2\pi 0/8} + x_1 \cdot e^{-j2\pi 5/8} + x_2 \cdot e^{-j2\pi 10/8} + x_3 \cdot e^{-j2\pi 15/8}$$
$$+ x_4 \cdot e^{-j2\pi 20/8} + x_5 \cdot e^{-j2\pi 25/8} + x_6 \cdot e^{-j2\pi 30/8} + x_7 \cdot e^{-j2\pi 35/8}$$

$$X_6 = x_0 \cdot e^{-j2\pi 0/8} + x_1 \cdot e^{-j2\pi 6/8} + x_2 \cdot e^{-j2\pi 12/8} + x_3 \cdot e^{-j2\pi 18/8}$$
$$+ x_4 \cdot e^{-j2\pi 24/8} + x_5 \cdot e^{-j2\pi 30/8} + x_6 \cdot e^{-j2\pi 36/8} + x_7 \cdot e^{-j2\pi 42/8}$$

$$X_7 = x_0 \cdot e^{-j2\pi 0/8} + x_1 \cdot e^{-j2\pi 7/8} + x_2 \cdot e^{-j2\pi 14/8} + x_3 \cdot e^{-j2\pi 21/8}$$
$$+ x_4 \cdot e^{-j2\pi 28/8} + x_5 \cdot e^{-j2\pi 35/8} + x_6 \cdot e^{-j2\pi 42/8} + x_7 \cdot e^{-j2\pi 49/8}$$

The corresponding flow graph is shown in Figure 10.2.

Again, you are encouraged to try a few calculations and verify that the FFT flow graph gives the same results as the DFT equations. Now we can see why the FFT is effective in reducing the number of computations. Each time the FFT doubles in size (N increases by a factor of 2), we need to add one more stage. For a 4-point FFT, we require two stages. For an 8-point FFT, we require three stages. For a 16-point FFT, four stages are required, and so on. The number of computations required for each stage is proportional to N. The required number of stages is equal to $\log_2$ N. Therefore, the FFT computational load increases by N $\cdot$ $\log_2$ N. The DFT computational load increases as N^2.

This is also the reason why FFT sizes are almost always in powers of 2 (2, 4, 8, 16, 32, 64, …). This is sometimes called a "radix 2" FFT. So rather than a 1000-point FFT, we see a 1024-point FFT. In practice, this common restriction to powers of 2 is not a problem.

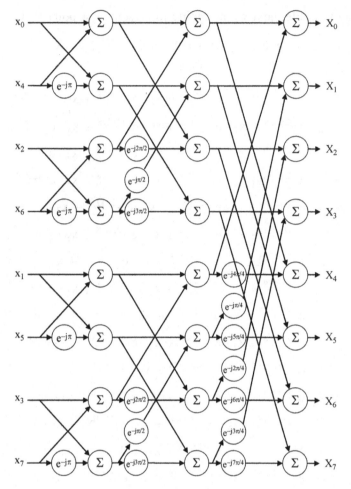

Figure 10.2

Another fairly common FFT implementation is the "radix 4" FFT. In this case, the butterfly has four complex inputs and four complex outputs. Although it does not reduce the number of operations, this type of FFT may sometimes be more efficient to implement in hardware or software. The FFT size would be restricted to powers of 4 in this case (4, 16, 64, 256, 2048, . . .).

10.3 Filtering Using the FFT and IFFT

The FFT algorithm is heavily used in many DSP applications. It is used whenever the signal needs to be processed in the spectral or frequency domain. Because it is so efficient to implement, sometimes even FIR filtering functions are performed using an FFT. This can be advantageous when the data is processed in batches. Rather than shifting the input data past a series of multipliers with filter coefficients, we can transform a buffer of N input data samples into the frequency domain using the FFT. This creates an N sample spectral representation of the input data. The coefficients represent the impulse, or time response, of the filter. This filter also has a corresponding frequency response, as discussed in earlier chapters. The spectral representation of the input created by the FFT can be multiplied by the filter frequency response, and the result then converted back to the time domain using the IFFT. This seems like a roundabout way, but this method often requires less work than traditional FIR filtering. In an FIR filter, each data sample must slide past all the coefficients, resulting in M multiplication operations per data sample in an M tap filter. This process is known as convolution. In the frequency domain, in contrast, the entire input spectral response is simply multiplied by the filter frequency response, which is only one multiplication operation per input data sample. This process is performed on groups of every N input data samples in a continuous manner. Even with the additional work of the FFT and IFFT, the net result still requires less computational effort than using an FIR filter, particularly when N is reasonably large, such as 1024.

10.4 Bit Growth in FFTs

As is apparent in the FFT flow diagrams in Figures 10.1 and 10.2, each stage of butterflies has complex exponential multiplication and summation operations. These operations are important when considering implementation. First, the multipliers are complex, and so is the data. This requires use of four real multipliers and two adder operations per complex multiplication, as shown here:

$$
\begin{aligned}
(A + jB) &\cdot (C + jD) \\
&= A \cdot C + jB \cdot C + A \cdot jD + jB \cdot jD \\
&= AC + jBC + jAD - BD \\
&= (AC - BD) + j(BC + AD)
\end{aligned}
$$

Second, the summation in each butterfly can result in a doubling of the signal amplitude if the two signals have the same phase. This process is known as FFT bit growth. It is 1 bit per stage in a radix 2 FFT and 2 bits per stage in a radix 4 FFT. This growth must be considered in fixed-point arithmetic. It can be most easily compensated for by simply shifting all the butterfly results right, or dividing by 2, before processing the next stage of butterflies. This operation ensures there can never be an overflow due to the summations. This is common when implementing fixed-point FFTs using a processor-based architecture. However, for small input signals, this truncation of the LSB at each stage can raise the quantization noise floor, leading to a loss in precision, especially in large FFTs.

In a hardware-based architecture, there is more flexibility. In particular, if the multiplier precision for one input operand can be increased, then the precision loss can be avoided.

Note that the complex exponential always has a magnitude of one, so the other operand of the multiplier is not required to grow. This is why sometimes asymmetrical-sized multipliers are used in FFT applications.

Another solution to this issue is to use floating-point arithmetic, which can allow for very large dynamic ranges. This usage is less common except in applications where the FFT precision (very low noise floor) is extremely important or when the size of the FFT is very high, due to the high implementation cost of floating point. For example, a 2^{20} point FFT (1 million points) would require double-precision floating point to implement the required precision in both the complex data and complex exponential coefficients.

10.5 Bit-Reversal Addressing

One last discussion point is bit-reversal. Notice the order of the x_i inputs to the 8-point FFT. This order is bit-reversed from normal sequential order (see the following table) to form the symmetry needed in the FFT structure. This bit-reversal addressing of the input order can be easily implemented in hardware by crossing, or reversing, the order of address bits. DSP processors can also perform this operation, usually by using a special bit-reversing addressing mode.

Table 10.4: Bit reversal example

Bit-Reversed Input (Decimal)	Bit-Reversed Input Index (Binary)	Sequential Input Index (Binary)	Sequential Input Index (Decimal)
0	000	000	0
4	100	001	1
2	010	010	2
6	110	011	3
1	001	100	4
5	101	101	5
3	011	110	6
7	111	111	7

In practice, the FFT can be set up to have either the input sequence bit-reversed or the output sequence bit-reversed (but not both). If the input is sequential, then the output is bit-reversed. If the input is bit-reversed (as in our examples), then the output is sequential. Either situation provides the needed symmetry. If the input (time domain) sequence is bit-reversed, this is called "decimation in time." If, on the other hand, it is chosen to have the output (frequency domain) sequence bit-reversed, then this is called "decimation in frequency." You may need to be aware of this vernacular when using FFT software or IP modules. Also, most DSP processors have special instructions that support bit-reversed addressing for use in FFT implementations.

CHAPTER 11

Digital Upconversion and Downconversion

Previously, we discussed complex modulation and baseband signals. The whole point of this discussion is to create a signal that can be used to carry the information bits from one location to another, whether over copper wire, a fiber-optic cable, or electromagnetically through the air. In nearly all cases, the signal needs to ride on a carrier frequency to be efficiently sent from one location to another. To do this, the signal frequency spectrum needs to be able to be moved up and down the frequency axis at will. This is a process of (frequency) upconversion and downconversion. All early methods used analog circuits to accomplish this task. In the past couple of decades, digital circuits, particularly FPGAs, have developed the computational capacity to perform these functions in many cases and offer important advantages over analog methods. These methods are known as digital upconversion and downconversion (also known as DUC and DDC, respectively, within the industry).

The process of upconversion is to take a signal that is at baseband (the frequency representation of the signal) and move or shift that frequency spectrum up to a carrier frequency. The width of the signal's frequency spectrum does not change; it is just moved to another part of the frequency spectrum. One common area of confusion is what happens to the negative part of the frequency spectrum. In a baseband signal, the negative frequency components overlie the positive components. Positive and negative frequencies are distinguished using complex representation. When the signal is upconverted, the positive and negative components are "unfolded" from on top of each, with the negative components below the carrier and the positive components above the carrier frequency.

A common example of this type of conversion occurs with speech. When you speak into a microphone, the sound waves create an electrical baseband signal, with frequency content from near 0 to about 3 kHz. With music, the frequency range can be much greater, up to 20 kHz. When we hear, the vibrations our ears detect are within this frequency range (in fact, our ears cannot hear frequencies beyond this range). This signal can be sampled and converted into a digital baseband signal, or it can remain as an analog signal. To transmit this signal over the air using electromagnetic radio waves, we need to increase the frequency. For example, the commercial AM radio system in the United States operates between 540 and 1600 kHz. The baseband speech signal is multiplied, or "mixed," with the much higher frequency carrier signal. This process superimposes the baseband signal on top of the carrier. For example, if the carrier frequency is 600 kHz, the upconverted 3 kHz speech signal will now ideally have a frequency or spectral

Digital Signal Processing 101. DOI: 10.1016/B978-1-85617-921-8.00015-8

content between $600 - 3$ kHz and $600 + 3$ kHz. This process also allows for multiple users to transmit and receive signals simultaneously because each user can be assigned a different carrier frequency and occupy different portions of the frequency spectrum. The upconversion and downconversion process can place the baseband signal at any desired "channel" frequency, allowing many different signals to occupy a common frequency band or range without interference. In the AM radio example, the carrier frequencies are spaced 10 kHz apart. This frequency separation is sufficient to prevent interference between stations.

Traditionally, this upconversion and downconversion process was done using analog signals and analog circuits. Analog upconversion with real (not complex) signals is depicted in Figure 11.1.

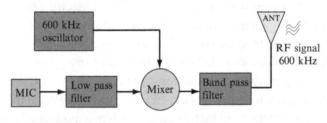

Figure 11.1

To see why a mixer works, let's consider a simple baseband signal of 1 kHz tone and a carrier frequency of 600 kHz. Each can be represented as a sinusoid. A real signal, such as a baseband cosine, has both positive and negative frequency components. At baseband, these components overlie each other, so this is not obvious. But once upconverted, the two components can be readily seen both above and below the carrier frequency.

The equation for the upconversion mixer is

$$\cos(\omega_{carrier}t) \cdot \cos(\omega_{signal}t) = \tfrac{1}{2} \cdot [\cos((\omega_{carrier} + \omega_{signal}) \cdot t) + \cos((\omega_{carrier} - \omega_{signal}) \cdot t)]$$

or

$$\cos(2\pi \cdot 600,000 \cdot t) \cdot \cos(2\pi \cdot 1000 \cdot t) = \tfrac{1}{2} \cdot [\cos(2\pi \cdot 599,000 \cdot t) + \cos(2\pi \cdot 601,000 \cdot t)]$$

The result is 2 tones, of half amplitude, at 599 and 601 kHz.

11.1 Digital Upconversion

This upconversion process can also be done digitally. Let's assume that the information content, whether it is voice, music, or data, is in a sampled digital form. In fact, as we covered in Chapter 9 on modulation, this digital signal is often in a complex constellation form, such as QPSK or QAM, for example.

If we want to transmit this information signal, at some point, it must be converted to the analog domain. In the past, the conversion from digital to analog occurred when the signal was in baseband form because the data converters could not handle higher frequencies. As the speeds and capabilities of analog-to-digital converters (ADCs) and digital-to-analog converters (DACs) improved, it became possible to perform the upconversion and downconversion digitally, using a digital carrier frequency. The upconverted signal, which has much higher frequency content, can then be converted to analog form using a high-speed DAC.

Upconversion is accomplished by multiplying the complex baseband signal (with I and Q quadrature signals) with a complex exponential of frequency equal to the desired carrier frequency.

The complex carrier sinusoid can be generated using a lookup table or implemented using any circuit capable of generating two sampled sinusoids offset by 90° (see Figure 11.3). If we want to do this digitally, the sample rates of the baseband and carrier sinusoid signal must be equal. Since the carrier signal is usually of much higher frequency than the baseband signal, the baseband signal has to be interpolated, or upsampled, to match the sample frequency of the carrier signal. Then the mixing, or upconversion, process results in the frequency spectrum shift depicted in Figure 11.2.

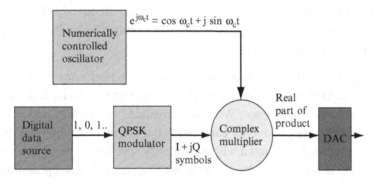

Figure 11.2

To simplify things, let's assume the output of the modulator is a complex sinusoid of 1 kHz. The equation for this upconversion process is

$$[\cos(\omega_{carrier}t) + j\sin(\omega_{carrier}t)] \cdot [\cos(2\pi \cdot 1000 \cdot t) + j\sin(2\pi \cdot 1000 \cdot t)] =$$
$$\tfrac{1}{2} \cdot [\cos((\omega_{carrier} - 2\pi \cdot 1000) \cdot t) + \cos((\omega_{carrier} + 2\pi \cdot 1000) \cdot t)] -$$
$$\tfrac{1}{2} \cdot [\cos((\omega_{carrier} - 2\pi \cdot 1000) \cdot t) - \cos((\omega_{carrier} + 2\pi \cdot 1000) \cdot t)] +$$
$$j \cdot \tfrac{1}{2} \cdot [\sin((\omega_{carrier} + 2\pi \cdot 1000) \cdot t) + \sin((\omega_{carrier} - 2\pi \cdot 1000) \cdot t)] +$$
$$j \cdot \tfrac{1}{2} \cdot [\sin((\omega_{carrier} + 2\pi \cdot 1000) \cdot t) + \sin((-\omega_{carrier} + 2\pi \cdot 1000) \cdot t)]$$

Simplifying, this becomes

$$\cos((\omega_{carrier} + 2\pi \cdot 1000) \cdot t) + j \cdot \sin((\omega_{carrier} + 2\pi \cdot 1000) \cdot t)$$

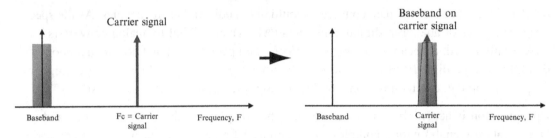

Figure 11.3

We can discard the imaginary portion. It is not needed because, at carrier frequencies, both positive and negative baseband frequency components can be represented by the spectrum above and below the carrier frequency.

The final result at the output of the DAC is

$$\cos\left((\omega_{carrier} + 2\pi \cdot 1000) \cdot t\right)$$

Note there is only a frequency component above the carrier frequency. The reason is that the input was a complex sinusiod rotating in the positive (counterclockwise) direction; there was no negative frequency component in this baseband signal.

Normally, a baseband signal has both positive and negative frequency components. For example, the complex QPSK modulator output can jump both clockwise and counterclockwise depending on the input data sequence. When upconverted, the positive and negative baseband components no longer overlie each other, but are unfolded on either side of the carrier frequency. The baseband signal with a frequency spectrum of 0-10 kHz occupies a total of 20 kHz, with 10 kHz on either side of the carrier frequency. But keep in mind that the baseband signal represents the positive and negative frequencies using quadrature form, so it is in the form of I and Q signals, each with a frequency spectrum from 0 to 10 kHz.

Frequently, there are several steps in upconverting to the final frequency used for transmission, as shown in Figure 11.4. There are several advantages to upconverting in steps. Generally, the DAC must operate at least 2½ times the carrier frequency. For signals in the gigahertz range, this exceeds the capacity of most DACs (although very high performance DACs can now operate at several gigahertz conversion rates). Another consideration is filtering. If several upconverting steps are used, filtering can be applied at each step.

Often, multiple stages of filtering are required to meet overall system requirements, such as the spectral emission requirements in the transmitted RF signal. These requirements are designed to ensure the transmitted signal does not cause interference with other users

A typical digital upconverting circuit is shown in Figure 11.4.

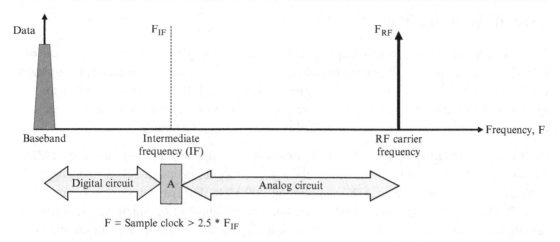

$$F = \text{Sample clock} > 2.5 * F_{IF}$$

Figure 11.4

11.2 Digital Downconversion

Digital downconversion is the opposite of upconversion. The circuit diagram for this process looks similar, as you can see in Figure 11.5.

Digital downconversion involves sampling, so naturally aliasing needs to be considered. The Nyquist sampling rule states that the sampling frequency must be at least twice the highest frequency of the signal being sampled. In practice, usually the sampling frequency is at least 2½ times the rate of signal frequency, to allow an extra margin for the transition band of the digital low-pass filters following.

But this is not quite true. It is possible to sample a signal at a frequency lower than its carrier frequency. The Nyquist rule applies to the actual bandwidth of the signal, not the frequency of the carrier.

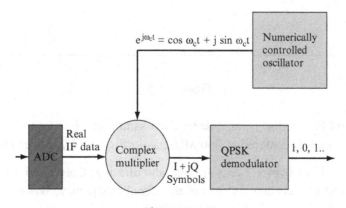

Figure 11.5

11.3 IF Subsampling

Using a technique called IF subsampling, we are able to sample at a much lower frequency than the carrier frequency. The term *IF subsampling* is used because the frequencies typically used in this technique lie somewhere between baseband and RF. The term *IF* refers to intermediate frequency. In this case, we are going to deliberately take advantage of an alias of the signal of interest. This is best illustrated using an example.

Let's use a 4G (fourth generation) wireless example. The signal of interest lies at 2500 MHz. Analog circuits are used to downconvert the signal to an IF, or intermediate frequency. Let's assume that we have an ADC sampling at 200 MHz. Further, let us also assume we have 20 MHz BW IF signal centered at 60 MHz, as in Figure 11.6, which we are trying to sample and downconvert for baseband processing. By the Nyquist rule, we can sample up to ½ the sample rate, or 100 MHz. To allow easier post-sampling filtering, we may want to limit this amount to 80 MHz instead. Our signal here lies between 50 and 70 MHz, so these conditions are met.

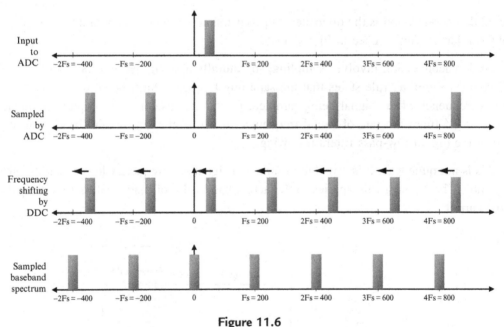

Figure 11.6

This process should be familiar from Chapter 3 on sampling. Now let us consider what happens if the IF signal is centered at 460 MHz, as in Figure 11.7, rather than 60 MHz.

As we also discussed in Chapter 3 on sampling and aliasing, there is no way to distinguish between a signal and an alias at a multiple of the sample frequency. When IF subsampling is performed, this aliasing can be used to our advantage.

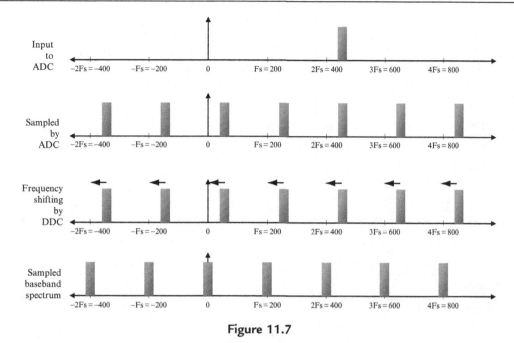

Figure 11.7

In our examples, any signal that aliases to a baseband frequency from −80 MHz to +80 MHz can work. Remember, the digital downconversion multiplies by a complex exponential, which can be rotating either counterclockwise or negative, thereby either shifting the spectrum left or right. Figure 11.8 provides another example, showing downconversion from a 340-MHz carrier frequency.

The signal at 340 MHz is aliased as if it were at −60 MHz and then can be shifted to baseband. The only areas of the spectrum that cannot be properly aliased down are those that alias to the region near the Nyquist frequency—in our case, 100 MHz. So signals close to 300 MHz, 500 MHz, and 700 MHz cannot be sampled in this way. But the rest of the spectrum can be sampled, so long as the sampling frequency meets the requirement of being more than twice the bandwidth of the signal of interest. In practice, the sampling frequency is often much higher than the signal bandwidth, which offers the additional advantage of allowing an increase in signal-to-noise ratio when the baseband signal is decimated to a lower frequency for baseband processing.

In our example, the downconverted signal has a spectrum from −10 to +10 MHz. This requires a sampling frequency of at least 20 MHz for both I and Q. If a decimate-by-4 FIR filter is used to low-pass-filter the downconverted signal, then further complex symbol processing can take place at 25 MHz. If we sample at a much higher rate and then low-pass-filter, a much greater percentage of the sampling quantization noise can be eliminated. Quantization noise, as discussed in Chapter 3, is due to the effect of the

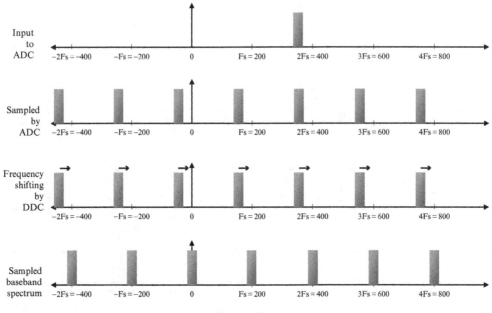

Figure 11.8

signal being sampled and mapped to specific amplitude levels that are limited by how many bits of resolution are available in the ADC. This noise is broadband, meaning that it is distributed evenly across the frequency spectrum. The effect of the low-pass filter is to attenuate much of this quantization sampling noise, along with unwanted signals, in the stopband region. The net effect is the equivalent of adding 1 bit of precision or 6 dB to the SNR, for every factor of two in decimation and filtering.

In our example, if we use a 12-bit ADC at 200 MHz F_s and perform the digital downconversion, low-pass-filter, and then decimate by 4, it is the equivalent of sampling at 50 MHz F_s using a 14-bit ADC. The decimation filter has, in essence, added 2 bits of resolution to the ADC, or 12 dB to the SNR of the sampled signal. This concept can be taken to extremes. There is a class of ADCs, called sigma-delta converters, that run at very high frequencies relative to the signal they are sampling, but use only a 1-bit sampler. An effective 10- or 12-bit ADC can be built using a single-bit sampling front end running at extremely high frequencies followed by decimation filtering stages.

Now let us go back to IF subsampling. In theory, we could sample a signal at any arbitrary frequency, but there must be some practical limitations. So far, we have discussed only the limitation of *not* sampling signals that alias to near the Nyquist frequency. But there are other limitations, and we discuss two of them here.

First, we must consider the performance of the ADC. Two principal characteristics are of concern here. Of course, the first is the maximum sampling rate at which the ADC operates.

In our example, we assumed an ADC that could sample at 200 MHz. When we use IF subsampling, we must also consider the analog bandwidth of the sampling circuit in the ADC. Ideally, the ADC samples the signal for an infinitely small instant of time and converts that measurement into a digital number. In practice, this circuit has a sampling window, or period of time in which it samples. The narrower this window, the higher signal frequencies it can sample. This specification is given in the datasheets provided by ADC manufacturers. In our example, we sampled a signal at 460 MHz. So we should check that our ADC has an analog front-end bandwidth of 500 MHz or higher.

Another factor that must be considered is clock jitter. Clock jitter is the amount of timing variability of the edge of the ADC sampling clock from cycle to cycle. It can be readily seen on an oscilloscope of sufficient quality. It is more easily seen as clock phase noise on a spectrum analyzer. Jitter shows up as spectral noise tapering off on either side of the clock frequency component. The less jitter, the more closely the clock appears as simply a vertical line in the frequency response. The effect of clock jitter is proportional to the frequency of the signal being sampled. It limits the SNR according to the following relationship:

$$SNR_{jitter} = 20 \log \left(1/(2\pi \cdot F_{signal} \cdot t_{jitter}) \right)$$

An example may make this concept clearer. In our example, the ADC clock is 200 MHz, which has a period of 5 nanoseconds (ns). Let us assume this clock has jitter of 5 picoseconds (ps). This doesn't sound like too much—only 0.1% of the clock period. First, we compute the SNR limitation due to clock jitter with an input signal at 60 MHz and then at 460 MHz:

60 MHz IF signal:

$$SNR_{jitter} = 20 \log \left(1/(2\pi \cdot 60 \cdot 10^6 \cdot 5 \cdot 10^{-12}) \right) = 54.5 \text{ dB}$$

460 MHz IF signal:

$$SNR_{jitter} = 20 \log \left(1/(2\pi \cdot 460 \cdot 10^6 \cdot 5 \cdot 10^{-12}) \right) = 36.8 \text{ dB}$$

This level of clock jitter limits the ADC performance for a 60 MHz signal at any level of precision beyond 10 bits. For a 460 MHz signal, the clock jitter limits the SNR to 36 dB, or about 6 bits. Clearly, this level of clock jitter is excessive in this IF subsampling application.

An analogy can be the same strobe light concept used in Chapter 3 on sampling and aliasing. The strobe light is assumed to flash at exact intervals. Any variance, or jitter, in the flashing intervals causes distortion in the position of the red dot on the wheel. The IF subsampling process would be as if the wheel made multiple rotations between strobe flashes, though it would appear that the wheel was rotating at a slower speed. But because the wheel is actually rotating much faster, any sampling errors due to jitter in the strobe flash are magnified, making the red dot appear blurred.

A very reasonable question to ask is if we can do something similar with a DAC to utilize aliased versions of the digitally upconverted signal. The answer is yes, but with an important limitation that discourages use of this technique in practice.

A major difference between an ADC and DAC is that while an ADC converts the signal to a digital representation, the DAC converts the digital samples to analog form *and* performs a "sample and hold" function of the analog signal between clocks. It is this sample and hold function that is a critical difference.

The sample and hold function is a rectangular filter in the time domain. Each digital sample input is an impulse of a given magnitude. The output is a rectangular shape of the input magnitude. The DAC impulse response, just like in any filter, defines the frequency response. In this case, the rectangular impulse response yields a sinc function [sin(x)/x] response in the frequency domain. This is shown in Figure 11.9, with the nulls in the frequency response corresponding to multiples of the DAC conversion or clock frequency (F_s).

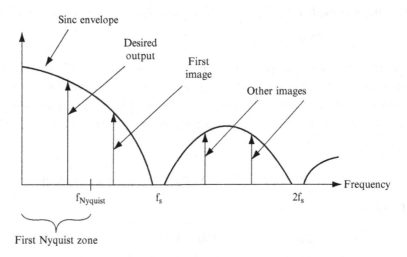

Figure 11.9

What this means in practice is that there will be a reduction in signal level at higher frequencies. By the time the DAC frequency response reaches the Nyquist frequency (½ F_s), it will have a droop of nearly 4 dB. The peak of the first lobe is 6 dB below the DC response, and the second lobe peak a further 6 dB lower. Due to this attenuation, usefulness of the DAC output at frequencies well above the DAC F_s is limited. Moreover, because the frequency response is not flat, it often needs to be compensated for. The closer the IF signal lies to multiples of the DAC F_s, the more distorted the frequency response is. Therefore, the IF frequency is usually limited to the first Nyquist zone. Furthermore, it is usually limited to about 80% of the Nyquist frequency or to 40% of the DAC F_s. For a DAC being clocked at 250 MHz, the upconverted IF signal should be at 100 MHz or less. For example, if the

IF carrier is chosen to be 70 MHz with a complex baseband signal extending to 10 MHz, the IF signal will occupy a spectrum from 60 to 80 MHz. After the DAC, the analog output will need to be filtered using an analog filter to remove the higher frequency DAC images and the DAC clock harmonics. Often, a surface acoustic wave (SAW) bandpass filter is used, and the choice of IF frequency is often determined in part by the frequencies where the SAW filters are commercially available.

Many systems, particularly if the IF signals have fairly wide bandwidth (like our 4G wireless example with an IF signal BW of 20 MHz), need to compensate for the frequency response droop due to the sinc response of the DAC even at these low IF frequencies. To do this, these systems use a sinc compensation FIR filter, after the NCO and complex multiplier performing the digital upconversion.

Figure 11.10 shows the frequency response of a typical sinc compensation FIR filter, calculated using a popular digital signal processing design tool called MATLAB® (available from The MathWorks). It compensates up to 80% of the Nyquist frequency. This digital filter would immediately precede the DAC and be placed after the digital upconversion process. It would have a data throughput rate equal to the DAC conversion rate. The associated coefficients are shown in Figure 11.11.

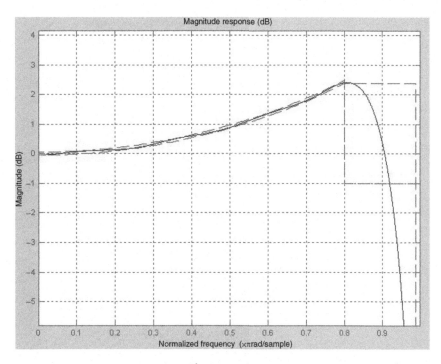

Figure 11.10

```
Numerator:
-0.0065468956647895373
 0.014409998368132634
-0.02915111437169457
 0.053017946956878462
-0.091122876704257547
 0.15519455343211477
-0.28282698843194176
 0.68492126102237871
 0.68492126102237871
-0.28282698843194176
 0.15519455343211477
-0.091122876704257547
 0.053017946956878462
-0.02915111437169457
 0.014409998368132634
-0.0065468956647895373
```

Figure 11.11
Sinc filter coefficients

CHAPTER 12

Error Correction Coding

This chapter provides an introduction to error correction coding and decoding. This process is also known as forward error correction, or FEC. This field is very complex and even has its own set of mathematics. A thorough understanding also requires background in probability and statistics theory, which is necessary to quantify the performance of any given coding method. Most texts on this topic delve into this math to some degree. Appendix E on binary field arithmetic gives a very quick, very basic, summary of the little Boolean arithmetic we use in this chapter.

As for the approach here, we skip nearly all mathematics and try to give an intuitive feel for the basics of coding and decoding. This is mostly done by example, using two types of error correcting codes. The first is a linear block code, specifically a Hamming code. The second is a cyclic code, using convolutional coding and Viterbi decoding.

To correct errors, we need to have a basic idea of how and why errors occur. One of the most fundamental mechanisms causing errors is noise. Noise is an unavoidable artifact in electronic circuits and in many of the transmission mediums used to transmit information. For example, in wireless and radar systems, there is a certain noise level present in the transmit circuitry, in the receive circuitry, and in the frequency spectrum used for transmission. During transmission, the transmitted signal may be attenuated to less than one billionth of the original power level by the time it arrives at the receive antenna. Therefore, small levels of noise can cause significant errors. There can also be interfering sources as well, but because noise is random, it is much easier to model and is used as the basis for defining most coding performance.

It makes sense that, given the presence of noise, we can improve our error rate by just transmitting at a higher power level. Since the signal power increases but the noise does not, the result should be better performance. This also leads to the concept of energy per bit, or E_{bit}. This measurement is computed by dividing the total signal power by the rate of information carrying bits.

One characteristic of all error correction methods is that we need to transmit more bits, or redundant bits, compared to the original bit rate being used. Consider a really basic error correction method. Suppose we just repeat each bit three times and have

Digital Signal Processing 101. DOI: 10.1016/B978-1-85617-921-8.00016-X the receiver take a majority vote across each set of three bits to determine what was originally transmitted. If one bit is corrupted by noise, the presence of the other two still allows the receiver to make the correct decision. But the trade-off is that we have to transmit three times more bits. Now assume we have a transmitter of a fixed power (say 10 watts). If we are transmitting three times more bits, that means the energy per bit is only one third what it was before, so each bit is now more susceptible to being corrupted by noise. To measure the effectiveness of a coding scheme, we can define a concept of coding gain. The goal is that with the addition of an error correcting code, the result is better system performance, and we want to equate the improvement in error rate due to error corrective coding to the equivalent improvement we could achieve if instead we transmitted the signal at a higher power level. We can measure the increase in transmit power needed to achieve the same improvement in errors as a ratio of new transmit power divided by previous transmit power and express it in dB_{power} (see Section 3.2, "Quantization," in Chapter 3 for a further explanation of decibels). This measurement is called the coding gain. The coding gain takes into account that redundant bits must be transmitted for any code, which actually reduces the signal transmit power per bit because the overall transmit power must be divided by the bit rate to determine the actual energy per bit available.

You can probably see that the basic idea is to correct errors in an efficient manner. This means being able to correct as many errors as possible using the minimum redundant bits. The efficiency is generally measured by coding gain. Again, the definition of coding gain is how much the transmit power of an uncoded system (in dB) must be increased to match the performance of the coded system. Matching performance means that the average bit error rates (BERs) are equal. Coding performance can also change at different BER rates, so sometimes this is expressed in graphical form.

Codes are often described by the parameters "k" and "n." The number of information bits in a codeword is given by k, and the total number of bits in a codeword, including the redundant bits, is given by n. The code rate is equal to k/n, and it tells us how much the data rate must be increased for a given code. Next, we are going to use a sample Hamming linear block code where k = 4 and n = 7.

12.1 Linear Block Encoding

With k information bits, we have 2^k possible input sequences. Specifically, with k = 4, we have 16 possible codewords. We could use a lookup table, but there is a more efficient method. Because this is a linear code, we need to map only the codewords generated by each of the k bits. For k = 4, we need to map only 4 inputs to codewords and then use the linear properties to build the remaining possible 12 codewords.

Inputs:

$$0001, 0010, 0100, 1000$$

Code mapping rule:

Map 4 input bits to bits 6–3 of a 7-bit codeword.

Then create redundant parity bits according to the following rules:

$$\text{bit } 2 = \text{bit } 6 + \text{bit } 5 + \text{bit } 3$$
$$\text{bit } 1 = \text{bit } 6 + \text{bit } 4 + \text{bit } 3$$
$$\text{bit } 0 = \text{bit } 5 + \text{bit } 4 + \text{bit } 3$$

Let's take the input 0001 (bit 3 of codeword = 1). Therefore, 0001 → 0001111.

With our 4 inputs, we get the following codewords:

$$1000 \rightarrow 1000110$$
$$0100 \rightarrow 0100101$$
$$0010 \rightarrow 0010011$$
$$0001 \rightarrow 0001111$$

Due to the linear property of the code, this is sufficient to define the mapping of all possible 2^k, or 16, input sequences. This process can be easily performed by a matrix:

$$[\text{7-bit codeword}] = [\text{4-bit input sequence}] \cdot \begin{bmatrix} 1000110 \\ 0100101 \\ 0010011 \\ 0001111 \end{bmatrix}$$

Notice this matrix, called a generator matrix **G**, is simply made up of the four codewords, where each row is one of the four single active bit input sequences. A simple example of generating the codeword follows:

$$[1101] \cdot \begin{bmatrix} 1000110 \\ 0100101 \\ 0010011 \\ 0001111 \end{bmatrix} = [1101100]$$

12.2 Linear Block Decoding

At the receiving end, we recover 7-bit codewords. There exists a total of 2^7, or 128, possible 7-bit sequences that we might receive. Only 2^4, or 16, of these are valid codewords. If we receive one of the 16 valid codewords, then there has been no error in transmission, and we can easily know what 4-bit input sequence was used to generate the codeword at the transmit end. But when an error does occur, we receive one of the remaining 112 codewords. We need a method to determine what error occurred and what the original 4-bit input sequence was.

To do this, we use another matrix, called a parity check matrix **H**. Let us look again at how the parity bit 2 is formed in this code, using the original parity definition:

$$\text{bit } 2 = \text{bit } 6 + \text{bit } 5 + \text{bit } 3$$

Based on this relationship, bit 2 = 1 only when bit 6 + bit 5 + bit 3 = 1. Given the rules of binary arithmetic, we can say bit 2 + bit 6 + bit 5 + bit 3 = 0 for any valid codeword. Extending this arithmetic, we can say this about any valid codeword:

bit 6 + bit 5 + bit 3 + bit 2 = 0	from parity bit 2 rule
bit 6 + bit 4 + bit 3 + bit 1 = 0	from parity bit 1 rule
bit 5 + bit 4 + bit 3 + bit 0 = 0	from parity bit 0 rule

This arithmetic comes from the earlier definition of the parity bit relationships we used to form the generator matrix. We can represent the preceding three equations in the form of parity generation matrix **H**.

$$\mathbf{H} = \begin{bmatrix} 1101100 \\ 1011010 \\ 0111001 \end{bmatrix}$$

As a quick arithmetic check, $\mathbf{G} \cdot \mathbf{H}^T$ must equal zero.

We can use the parity check matrix **H** to decode the received codeword. We can compute something called a syndrome **S** as follows:

$$[\text{received codeword}] \cdot \mathbf{H}^T = \mathbf{S} \text{ (the syndrome)}$$

When the syndrome is zero, all three parity equations are satisfied, there is no error, a valid codeword was received, and we can simply strip away the redundant parity bits from the received codeword, thus recovering the input bits. When the syndrome is non-zero, an error has occurred. Because this is a linear code, we can consider any received codeword to be the sum of a valid codeword and an error vector or word. Errors occur whenever the error vector is nonzero. This result leads to the realization that the syndrome depends only on the error vector, and not on the valid codeword originally transmitted.

In our example, the syndrome is a 3-bit word, capable of representing 8 states. When $\mathbf{S} = [0,0,0]$, no error has occurred. When the syndrome is any other value, an error has occurred, and the syndrome indicates in which bit position the error is located. But which syndrome value maps to which 7 possible error positions in the received codeword?

Let's go back to the parity relationship that exists when S = 0:

$$\text{bit } 6 = \text{bit } 5 + \text{bit } 3 + \text{bit } 2 = 0$$
$$\text{bit } 6 = \text{bit } 4 + \text{bit } 3 + \text{bit } 1 = 0$$
$$\text{bit } 5 = \text{bit } 4 + \text{bit } 3 + \text{bit } 0 = 0$$

Let's examine our earlier example of the valid codeword [1101100]. The parity check matrix is

$$\mathbf{H} = \begin{bmatrix} 1101100 \\ 1011010 \\ 0111001 \end{bmatrix} \quad \text{and} \quad \mathbf{H}^T = \begin{bmatrix} 110 \\ 101 \\ 011 \\ 111 \\ 100 \\ 010 \\ 001 \end{bmatrix}$$

$$\mathbf{S} = \begin{bmatrix} 1101100 \end{bmatrix} \cdot \begin{bmatrix} 110 \\ 101 \\ 011 \\ 111 \\ 100 \\ 010 \\ 001 \end{bmatrix} = \begin{bmatrix} 0,0,0 \end{bmatrix}$$

By inspection of the parity equations, we can see that syndromes with only one nonzero bit must correspond to an error in the parity bits of the received codeword.

S = [1,0,0] indicates the first parity equation is not satisfied, but the second and third are true. By inspection of parity bit definitions, we can see this must be caused by bit 2 of the received codewords because this bit alone is used to create bit 2 of the syndrome. Similarly, an error in bit 1 in the received codeword creates the syndrome [0,1,0], and an error in bit 0 in the received codeword creates the syndrome [0,0,1].

An error in bit 3 of the received codeword causes a syndrome of [1,1,1] because this bit appears in all three parity equations. Therefore, we can map the syndrome value to specific bit positions where an error has occurred:

S = [0,0,0] → no error in received codeword
S = [0,0,1] → error in bit 0 of received codeword
S = [0,1,0] → error in bit 1 of received codeword
S = [0,1,1] → error in bit 4 of received codeword
S = [1,0,0] → error in bit 2 of received codeword

S = [1,0,1] → error in bit 5 of received codeword
S = [1,1,0] → error in bit 6 of received codeword
S = [1,1,1] → error in bit 3 of received codeword

The procedure to correct errors is to multiply the received codeword by $\mathbf{H}^T$, which gives $\mathbf{S}$. The value of $\mathbf{S}$ indicates the position of error in the received codeword to be corrected. This concept can best be illustrated with an example.

Suppose we receive the codeword [1111100]. We then can calculate the syndrome as follows:

$$S = [1111100] \cdot \begin{bmatrix} 110 \\ 101 \\ 011 \\ 111 \\ 100 \\ 010 \\ 001 \end{bmatrix} = [0,1,1]$$

This syndrome indicates an error in bit 4. The corrected codeword is [1101100], and the original 4-bit input sequence 1,1,0,1, which is not the 1,1,1,1 of the received codeword.

12.3 Minimum Coding Distance

A good question is what happens when there are two errors simultaneously. Hamming codes can detect and correct only one error per received codeword. The amount of detection and correction a code can perform is related to something called the minimum distance. For Hamming codes, the minimum distance is three. This means that all the transmitted codewords have at least 3 bits different from all the other possible codewords. Recall that in our case, we have 16 valid codewords out of 128 possible sequences. They are as follows:

0000000	0001111	0010011	0011100
0100101	0101010	0110110	0111001
1000110	1001001	1010101	1011010
1100011	1101100	1110000	1111111

Each of these codewords has 3 or more bit differences from the other 15 codewords. In other words, each codeword has a minimum distance of three from neighboring codewords. That is why with a single error, it is still possible to correctly find the closest codeword, with respect to bit differences. With two errors, the codeword is closer to the wrong codeword, again with respect to bit differences. This is analogous to trying to map a symbol in two-dimensional I-Q space to the nearest constellation point in a QAM demodulator. There are other coding methods with greater minimum distances, able to correct multiple errors. We look at one of these next.

12.4 Convolutional Encoding

A second major class of channel codes is known as convolutional codes. Convolutional codes can operate on a continuous string of data, whereas block codes operate on words. Convolutional codes also have memory; the behavior of the code depends on previous data.

Convolutional coding is implemented using shift registers with feedback paths. There is a ratio of "k" input bits to "n" output bits, as well as a constraint length "K." The code rate is k/n. The constraint length K corresponds to the length of the shift register and also determines the length of time or memory that the current behavior depends on past inputs.

Next, let's go through a very simple convolutional coding and Viterbi decoding example. We use an example with $k = 1$, $n = 2$, and $K = 3$. The encoder is described by generator equations, using polynomial expressions to describe the linear shift register relationships. The first register connection to the XOR gate is indicated by the 1 in the equations, the second by X, the third by X^2, and so forth. Most convolution codes have a constraint length less than 10.

The rate is defined as the ratio of input bits to output bits. The rate here is ½ as there are every input bit M, there are two output bits N.

The output is usually interleaved in a bitstream $N1_j$, $N2_j$, $N1_{j+1}$, $N2_{j+1}$, $N1_{j+3}$, $N2_{j+3}$, ... (See Figure 12.1.)

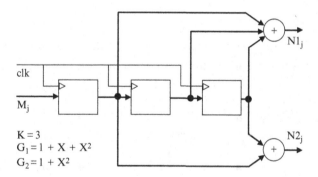

Figure 12.1

The following table shows encoder operation with the input sequence 1,0,1,1,0,0,1,0,1,1,0,1 for each clock cycle. The register is initialized to zero.

The resulting output sequence is as follows:

$$\{1,1\}\{1,0\}\{0,0\}\{0,1\}\{0,1\}\{1,1\}\{1,1\}\{1,0\}\{0,0\}\{0,1\}\{0,1\}\{0,0\}$$

Notice that the output $N1_j$ and $N2_j$ are both a function of the input bit M_j and the two previous input bits M_{j-1} and M_{j-2}. The previous $K - 1$ bits, M_{j-1} and M_{j-2}, form the state of the

Table 12.1: Encoder state and outputs

Register Value	N1 Value	N2 Value	Time or Clock Value
1 0 0	1	1	T_1
0 1 0	1	0	T_2
1 0 1	0	0	T_3
1 1 0	0	1	T_4
0 1 1	0	1	T_5
0 0 1	1	1	T_6
1 0 0	1	1	T_7
0 1 0	1	0	T_8
1 0 1	0	0	T_9
1 1 0	0	1	T_{10}
0 1 1	0	1	T_{11}
1 0 1	0	0	T_{12}

encoder state diagram. As shown in Figure 12.2, the encoder is at a given state. The input bit M_j causes a transition to another state at each clock edge, or T_j. Each state transition results in an output bit pair $N1_j$ and $N2_j$. Only certain state transitions are possible. Transitions due to a 0-bit input are shown in dashed lines, and transitions due to a 1-bit input are shown in solid lines. The output bits shown at each transition are labeled on each transition arrow in the figure.

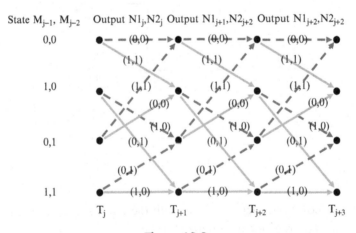

Figure 12.2

Now, let us trace the path of the input sequence through the trellis using Figure 12.3 and the resulting output sequence. This task helps us gain the insight that the trellis is representative of the encoder circuit because use of the trellis will be key in Viterbi decoding. The highlighted lines show the path of input sequence.

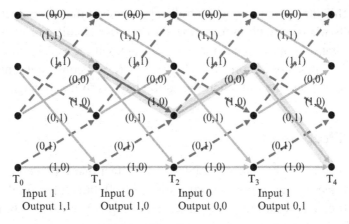

Input 1   Input 0   Input 0   Input 1
Output 1,1   Output 1,0   Output 0,0   Output 0,1

Figure 12.3-1

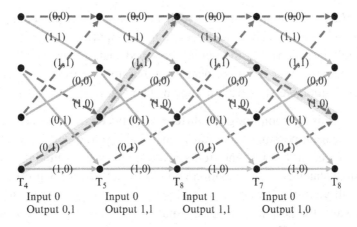

Input 0   Input 0   Input 1   Input 0
Output 0,1   Output 1,1   Output 1,1   Output 1,0

Figure 12.3-2

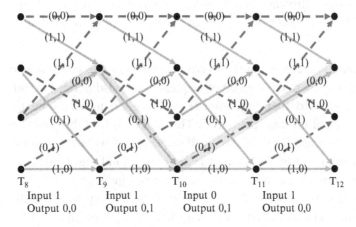

Input 1   Input 1   Input 0   Input 1
Output 0,0   Output 0,1   Output 0,1   Output 0,0

Figure 12.3-3

By tracing the highlighted path through the trellis, you can see that the output sequence is the same as our results when computing using the shift register circuit. For constraint length K, we have $(K - 1)^2$ states in our trellis diagram. Therefore, with K = 3 in our design example, we have 4 possible states. For a more typical K = 6 or K = 7 constraint length, there would be 32 or 64 states, respectively, although this is too tedious to try to diagram.

12.5 Viterbi Decoding

The Viterbi decoding algorithm takes advantage of the fact that only certain paths through the trellis are possible. For example, starting from state 00, the output on the next transition must be either 0,0 or 1,1, resulting in the next state being either 0,0 or 1,0, respectively. The output of 1,0 or 0,1 is not possible, and if this sequence occurs, then an error must be present in the received bit sequence.

We first look at Viterbi decoding using the sequence given in the preceding section as our example. Keep in mind that when we are decoding, the input data is unknown (whether a "dashed" or "solid" transition), but the output (received) data is known. The job of the decoder is to correctly recover the input data using possibly corrupted received data. The decoder does assume we start from a known state (M_{j-1}, M_{j-2} = 0, 0).

We are going to do this by computing the difference between the received data pair and the transition output for each possible state transition. We keep track of this cost for each transition. Once we get further into the trellis, we can check the cumulative cost entering each of the possible states at each transition and eliminate the path with the higher cost. In the end, this will yield the lowest cost valid path (or valid path closest to our received sequence). As always, the best way to get a handle on this concept is through an example. First, we look at the cost differences with a correct received sequence (no errors) and then one with errors present. In the diagram shown in Figure 12.4, the figures on each transition arrow are the absolute difference (Δ) between the two received bits and the encoder output generated by that transition (as shown on the encoding trellis diagram above in Figure 12.3-1).

Notice that the cumulative Δ is equal to zero as we follow the path the encoder took (solid highlighted) to generate the received sequence. But also notice two other paths that can arrive at the same point (dashed highlighted) and have a nonzero cumulative Δ. The cumulative cost of each of these paths is 5. The idea is to find the path with the least cumulative Δ or difference from the received sequence. If we are decoding, we can then recover the input sequence by following the zero cost (solid highlighted) path. The dashed arrows indicate a "zero" input bit, and the solid arrows indicate a "one" input bit. The sequence recovered is therefore 1, 0, 1, 1. Notice that this sequence matches the first 4 bits input to the encoder. This is called "maximum likelihood decoding."

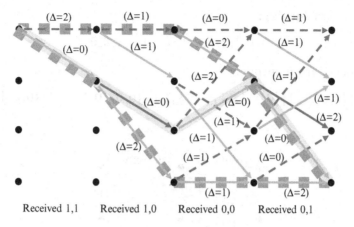

Figure 12.4

This approach is all well and good, but as the trellis extends further into time, there are too many possible paths merging and splitting to calculate the cumulative costs. Imagine having 32 possible states, rather than the 4 in our example, and a trellis extending for hundreds of transitions. What the Viterbi algorithm does is to remove, or prune, less likely paths. Whenever two paths enter the same state, the path having the lowest cost is chosen. The selection of lowest cost, or surviving path, is performed for all states at each transition (excluding some states at the beginning and end, when the trellis is either diverging or converging to a known state). The decoded path advances through the trellis, eliminating the higher cost paths and, at the end, will backtrack along the path of cumulative least cost to extract the most likely sequence of received bits. In this way, the codeword or valid sequence most closely matching the actual received sequence (which can be considered a valid code sequence plus errors) will be chosen.

Before proceeding further, we need to consider what happens at the end of the sequence. To backtrack after all the paths' costs have been computed and the less likely paths pruned, we need to start backtracking from a known state. To achieve this, $K - 1$ zeros are appended to the input bit sequence entering the encoder. This guarantees that the last state in the encoder, and therefore the decoder trellis, must be the zero state (the same state we start from, as the encoder shift register values are initialized to zero). So we add $K - 1$ or two zeros to our input sequence, shown in bold below, and use them in our Viterbi decoding example. The longer the bit sequence, the less impact the addition of $K - 1$ zeros has on the overall code rate because these extra bits do not carry information and have to be considered as part of the "redundant bits" (present only to facilitate error correction). So our code rate is slightly less than k/n. (See Figure 12.5.)

Input bit sequence to encoder:

$$1, 0, 1, 1, 0, 0, 1, 0, 1, 1, 0, 1, \mathbf{0}, \mathbf{0}$$

Output bit sequence from encoder:

$$\{1, 1\}\{1, 0\}\{0, 0\}\{0, 1\}\{0, 1\}\{1, 1\}\{1, 1\}\{1, 0\}\{0, 0\}\{0, 1\}\{0, 1\}\{0, 0\}\{\mathbf{1}, \mathbf{0}\}\{\mathbf{1}, \mathbf{1}\}$$

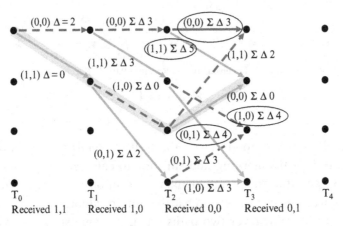

Figure 12.5

The circles in Figure 12.5 identify the four higher-cost paths that can be pruned at T_3, based on the higher cumulative path cost ($\Sigma\Delta$ = sum of delta or the difference from the received bit sequence). The results after pruning are shown in Figures 12.6 and 12.7.

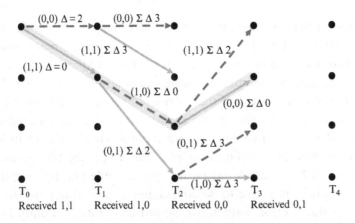

Figure 12.6

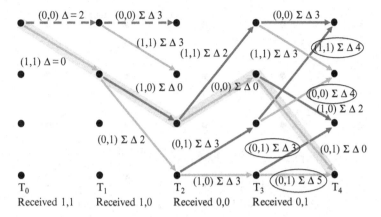

Figure 12.7

Next, the higher-cost paths (circled) are pruned at T_4, again eliminating ½ of the paths again. The path taken by the transmitter encoder is again highlighted at zero cumulative cost because there are no receive errors. (See Figures 12.8 and 12.9).

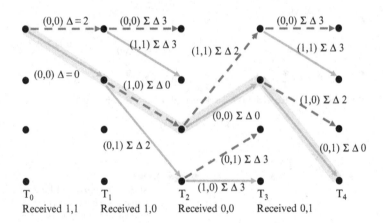

Figure 12.8

Due to the constant pruning of the high-cost paths, the cumulative costs are not higher than 3 so far. (See Figure 12.10.)

Next, we have the K − 1 added bits to bring the trellis path to state 0. (See Figure 12.11.)

We know that, due to the K − 1 zeros added to the end of the encoded sequence, we must start at state 0 at the end of the sequence, at T_{14}.

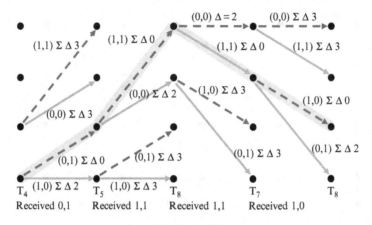

Figure 12.9

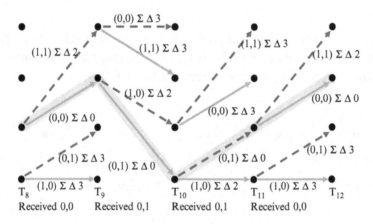

Figure 12.10

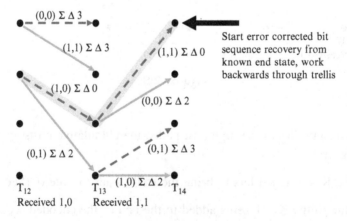

Start error corrected bit
sequence recovery from
known end state, work
backwards through trellis

Figure 12.11

We can simply follow the only surviving path from state 0,0 at T_{14} backward, as highlighted. We can determine the bit sequence by the use of dashed or solid lines for the arrows (dashed for 0, solid for 1). In a digital system, a 0 or 1 flag would be set for each of the four states at each T_j whenever a path is pruned, identifying the original bit M_j as 0 or 1 associated with the surviving transition.

Now let's reexamine what happens in the event of receive bit errors (indicated in bold in Figure 12.12). We assume that there is a bit error at transition T_8 and T_{10}. The correct path (generated by the encoder) through the trellis is again highlighted. Note how the cumulative costs change and the surviving paths change. However, in the end, the correct path is the sole surviving path that reaches the end point with the lowest cost.

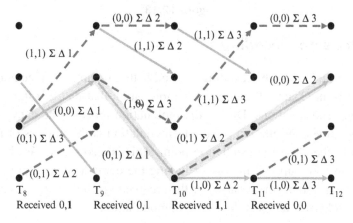

Figure 12.12

When we had no errors, we found all the competing paths had different costs, and we would always prune away the higher-cost path. However, in the presence of errors, we occasionally have paths of equal cost. In these cases, it does not really matter which path we choose. For example, we can just make an arbitrary rule and always prune away the bottom path when paths of equal cumulative costs merge. And if we use soft decision decoding, which is explained shortly, the chances of equal-cost paths is very low.

The path ending at state 0, at T_{14}, has a cumulative cost of 2 because we corrected two errors along this path (see Figure 12.13). Notice that some of the pruned paths are different from the no-error Viterbi decoding example, but the highlighted path that is selected is the same in both cases. The Viterbi algorithm finds the most likely valid encoder output sequence in the presence of receive errors and is able to do so efficiently because it prunes away less likely (higher cost) paths continuously at each state where two paths merge together.

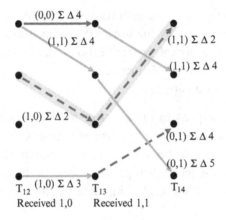

Figure 12.13

12.6 Soft Decision Decoding

Another advantage of the Viterbi decoding method is that it supports a technique called "soft decision." Recall from Chapter 9 on complex modulation that data is often modulated and demodulated using constellations. During the modulation process, the data is mapped to one of several possible constellation points. Demodulation is the process of mapping the received symbol to the closest constellation point. This is also known as "hard decision" demodulation. However, suppose instead of having the demodulator output the closet constellation point, it outputs the location of the received symbol relative to the constellation points. This is also known as "soft decision" demodulation.

For example, in the constellation in Figure 12.14, imagine that we receive the symbols labeled S_1, S_2, S_3, and S_4. We demodulate these symbols as shown in the following table.

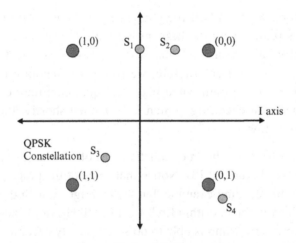

Figure 12.14

Table 12.2: Demodulator outputs

	S_1	S_2	S_3	S_4
Hard decision demodulation	0*, 0	0, 0	1, 1	0, 1
Soft decision demodulation	½, 0	¼, 0	¾, ¾	0, 1

*as equal distance between two points, arbitrarily assign to one of the points

Basically, we tell the Viterbi decoder how sure we are of the correct demodulation. Instead of giving a yes or no at each symbol, we can also say "maybe" or "pretty sure." The received signal value in the trellis calculation is now any value between 0 and 1, inclusive. This result is then factored into the path costs and the decision on which merged path to prune away.

Simulation and testing have shown over 2 dB improvement in coding gain when using soft decision (16 levels, or 4 bits) compared to hard decision (2 levels, or a single bit) representation for the decisions coming from the demodulator. The additional Viterbi decoding complexity for soft decoding is usually negligible.

There are many other error corrective codes besides the two simple ones we have presented here. Some common codes used in the industry are Reed Solomon, BCH, and Turbo decoding. Low density parity code (LDPC) is another emerging coding technique, which promises even higher performance at a cost of much increased computational requirements. In addition, different codes are sometime concatenated to further improve performance.

12.7 Cyclic Redundancy Check

A cyclic redundancy check (CRC) word is often appended at the end of a long string or packet of data, prior to the error correcting encoder. The CRC word is formed by taking all the data words and exclusive-ORing (EXOR) each word together. For example, the data sequence might be 1024 bits, organized as 64 words, each 16 bits. This would require 16-1 EXOR word operations in a DSP or in hardware to form the 16-bit CRC word. This function is analogous to that of a parity check bit for a single digital word being stored and accessed from dynamic random access memory (DRAM).

At the conclusion of the error decoding process, the recovered input stream can again be partitioned in words and EXORed together and the result compared to the recovered CRC word. If they match, then we can be assured with very high probability that the error correction was successful in correcting any error in transmission. If not, then there were too many errors to be corrected, and the whole data sequence or frame can be discarded. In some cases, there is a retransmission facility built into the higher-level communication protocol for these occurrences, and in others, such as a voice packet in a mobile phone system, the data is real time, so alternate mechanisms of dealing with corrupt or missing data are used.

12.8 Shannon Capacity and Limit Theorems

No discussion on coding should be concluded without at least a mention of the Shannon Capacity Theorem and Shannon limit. The Shannon Capacity Theorem defines the maximum amount of information, or data capacity, that can be sent over any channel or medium (wireless, coax, twister pair, fiber, etc.):

$$C = B \log_2(1 + S/N)$$

where

  C is the channel capacity in bits per second (or maximum rate of data).
  B is the bandwidth in Hz available for data transmission.
  S is the received signal power.
  N is the total channel noise power across bandwidth B.

What this theorem says is that the higher the signal-to-noise ratio and the more channel bandwidth, the higher the possible data rate. This equation sets the theoretical upper limit on data rate, which, of course, is not fully achieved in practice. This equation does not make any limitation on how low the achievable error rate will be. That is dependent on the coding method used.

As a consequence, the minimum SNR required for data transmission can be calculated. This is known as the Shannon limit, and it occurs as the available bandwidth goes to infinity:

$$E_b/N_0 = -1.6 \, \text{dB}$$

where

  E_b is the energy per bit.
  N_0 is the noise power density in Watts/Hz ($N = BN_0$).

If the E_b/N_0 falls below this level, no modulation method or error correction method will allow data transmission.

These relationships define maximum theoretical limits against which the performance of practical modulation and coding techniques can be compared. As newer coding methods are developed, we are able to get closer and closer to the Shannon limit, usually at the expense of higher complexity and computational rates.

CHAPTER 13

Analog and TDMA Wireless Communications

The first mobile phone systems were based on analog technology. Developed in the 1970s, this system in the United States was known as the American mobile phone system (AMPS). Similar systems were developed in Europe and Japan. These are known as first generation, or "1G," systems. What differentiated these systems from previous wireless systems (police or public service radio, citizens' band radio, military systems) was the concept of frequency reuse. Because these systems reused frequency channels, enough frequency was available that each pair of users could be assigned their own, private, communication link. Once the call was finished, this frequency channel could be assigned to other users. An important aspect of frequency reuse was that a large region of phone service or coverage could be divided into sections, or "cells," to allow the same frequency to be used over and over, providing these locations were not too close to each other. This was the origin of the name "cellular" phone service. This allowed a large pool of users to be serviced with a much smaller set of frequency channels, which allowed for efficient use of the frequency spectrum.

In the cellular diagram in Figure 13.1, each cell also has its own base station or transmitter/receiver equipment connected to a central network. The cells are lettered, depicting that each different letter cell is assigned a different set of frequency channels. This is a seven-cell reuse pattern. Each of the seven letters uses different frequencies, and no cells of the same letter are adjacent to each other. Additionally, separate frequency bands are used for downlink transmission (base station → mobile phones) and uplink (mobile phones → base station). This is known as frequency division duplex, or FDD. For example, the AMPS system uses the 824-849 MHz band for downlink and 869-894 MHz band for uplink. Each individual user channel occupies 30 kHz, resulting in 800 channel pairs. If we have frequency reuse every seven cells, then about 114 channel pairs are available in each cell. If we assume that at peak usage times, 10% of cell phone subscribers are making phone calls, then over 1000 users can be serviced by any given cell.

In addition, most base stations use directional panel antennas. There are typically one transmit and two receive antennas per sector. A sector is roughly 120°,

Digital Signal Processing 101. DOI: 10.1016/B978-1-85617-921-8.00017-1

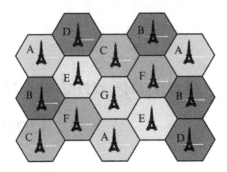

Figure 13.1

with some overlap. There are three sectors per cell site, and the base station antenna tower has each sector's antenna pointing in three different directions, 120° apart.

Each user is able to move about within the boundaries of a given sector of a given cell, which is defined by the RF coverage area of the base station within that sector. But to move further than this, or roam, requires a centralized network control system. When the mobile phone reaches the edge cell, it can detect the weakening strength of the base station signal. The base station can also detect the lower signal strength of that mobile handset. The mobile phone is instructed by the network via that base station to scan the frequencies of adjacent sectors and of neighboring base stations. When an adjacent sector or neighboring base station's signal is found to be stronger, then the network performs a handoff. The mobile phone is instructed to switch to an unoccupied frequency channel of the adjacent sector or neighboring base station. Simultaneously, the landline connection of the phone call in progress is switched from the original base station sector to the adjacent sector or to the neighboring base station. In this way, a phone call can be carried on continuously as the mobile phone travels throughout the network coverage area of the network, defined by the contiguous RF coverage of the network's or service provider's base stations.

13.1 Early Digital Innovations

Two technology breakthroughs made this type of digital service possible. For this type of network to function, the mobile phones have to be intelligent enough to receive and reply to commands, perform signal strength measurements, and rapidly tune to new frequencies on command. This required a low-power and low-cost microprocessor, which could be incorporated into the mobile phones. It also required something called a fractional "N" synthesizer, which was used to build a digitally controlled phase-locked loop oscillator circuit. This allowed for software-controlled frequency tuning, to one of many closely spaced frequency channels. Both of these technologies became available in the 1970s and were essential to the development of an intelligent, frequency agile mobile cell phone. However, the baseband and RF processing of the voice signal from microphone/speaker to the antenna in the mobile phone was, for the most part, implemented using traditional analog circuits and techniques.

In each cell, most of the frequency channels are available for use as voice channels that can be assigned to a given user during a call. But one or more of the channels in each cell are used as control channels. A phone that is not in conversation monitors one of these pre-assigned control channels. Here, it can monitor RF signal strength and listen for paging messages and other commands. The modulation used is called frequency shift keying, or FSK. It is like Morse code, except instead of dashes and dots, two different frequency tones are transmitted to indicate "ones" and "zeros." All mobile phones would reply to commands using a common uplink channel, known as a random access channel, or RACH. This was a somewhat uncontrolled process because several mobiles could transmit simultaneously on this channel. To ensure message replies would eventually get through, each phone would wait a random amount of time before retransmitting in the event of a collision. The great majority of channels were designated as voice channels, which carried the voice conversation using frequency modulation, similar to any FM radio station. In fact, in the early days of analog cell phones, a person could eavesdrop on wireless conversations using a simple FM scanner (but hear only one side of the conversation at a time because in an FDD system, there are separate uplink and downlink frequency bands).

13.2 Frequency Modulation

Information can be carried by a sinusoidal wave using varying amplitude, frequency, and phase. In QAM modulation, the amplitude and phase are changed. In FM modulation, only the frequency is modified. FM is a modulation method inherently suitable for an analog input or baseband signal. Basically, the instantaneous frequency of the carrier is made to increase or decrease from the carrier frequency by an amount proportional to the modulating or baseband signal. This change in the carrier frequency is known as the frequency deviation. The frequency deviation is proportional to the amplitude of the baseband input. The rate of change (derivative) of the carrier frequency is proportional to the frequency of the baseband input. The AMPS system used FM with a peak derivative of 12 kHz.

Since there is no amplitude modulation, the FM signal is of constant amplitude. This is the inherent FM characteristic that is superior over AM modulation, and why FM radio, from its beginnings in the 1930s, was designed for high fidelity compared to AM radio. Any additive noise with an AM signal causes distortion of the amplitude, which is the baseband signal. In contrast, with FM, the frequency carries the baseband signal and is much less affected by additive noise. This additive noise causes phase distortion, which can affect the frequency demodulation, but most of this can be filtered out of the resulting baseband signal. Another important characteristic of FM modulation is that, due to the constant amplitude characteristic, it can be very efficiently amplified. This topic is further discussed in chapter 15.

13.3 Digital Signal Processor

The analog-based cell phone system brought a landline-telephone-like experience to mobile communications. The invention of the digital signal processor, or DSP, paved the way for the next step in the evolution of wireless phone communication.

The DSP is basically a specialized microprocessor. It has at least one dedicated multiplier with an associated accumulator, or adder with a feedback path. This processor can be used to efficiently calculate a sum of products, used in FIR filters. DSPs, unlike most microprocessors, can fetch instruction and data words from memory simultaneously. To be able to do one calculation per clock cycle, the DSP generally requires at least three data buses. One data bus is used to fetch the instruction word, and two more buses to fetch the two operands for the multiplier from memory. Sometimes there is a fourth dedicated data bus to be able to simultaneously store data back to memory. This memory is generally all single-cycle access, on-chip, also known as Level 1 memory. In DSP, the data, unlike instructions, is usually read and written in a predictable manner. Therefore, DSPs contain at least two data address generators, which can be preconfigured to calculate addresses in a given pattern and even in a circular, or repeating, manner. This allows implementation of virtual shift registers in memory and accessing of filter coefficients in the correct order. There is often a "bit-reversing" mode, which can be used to read or write FFT data in a decimation in time or decimation in frequency fashion (refer to Chapter 10 on FFT for more detail).

In addition, DSP instructions often have data-shifting capabilities, which allow for the decimal point to be aligned as needed prior to saving data to memory. The data shifting can also be configured as a barrel shifter and, in conjunction with logical operations, used to implement many error coding operations. Given the popularity of the Viterbi algorithm, there is often a special instruction to implement path-metric comparisons and selections. Various accumulator data rounding and saturation modes are often supported. To obtain maximum performance, DSPs often had some instructions with pipeline restrictions, which created exception cases for the programmer.

DSPs were programmed using a manufacturer-specific assembly language, usually by firmware engineers extremely familiar with the details of the DSP hardware architecture. As a consequence, the majority of DSP programmers came not from a software programming background, but from an electrical engineering background. Due to the small on-chip memory available, the need to minimize the number of clock cycles per calculation, and the intricate, mathematical nature of the algorithms being implemented, DSP programming became as much an art as a skill. Current DSPs come with advanced C compilers, making code development much more efficient, although assembly macros are still used for specialized instructions because they often operate in a parallel manner.

With the advent of the first DSPs, it didn't take long for many applications to develop. Among the most important were digital mobile phone systems.

These systems were the second generation of mobile phone technology, now known as "2G." These systems are known as time division multiple access, or TDMA. This fancy term just means that multiple users rotate turns using both the uplink and downlink frequency channels, allowing more simultaneous users.

13.4 Digital Voice Phone Systems

A key feature of a digital mobile phone system is that the voice is digitized. Landline phone systems have long been digital. The actual phones in homes and businesses are analog and, using twisted-pair phone lines, connect to the local telephone exchange (this same line also carries the DSL signal for an Internet connection). At the telephone exchange, the voice signal is digitized using ADCs and DACs sampling at 8 kHz. The samples are not mapped linearly, but logarithmically, into an 8-bit digital representation, using a process known as companding. This technique reduces the quantization noise at low signal levels at the expense of quantization noise at higher signal levels, effectively resulting in a higher dynamic range. The voice signal is now in a digital format, with 8-bit samples at 8 kHz, for a resulting bit rate of 64 kilobits per second (kbps). In this form, the signal can be managed and transmitted by telephone switches and systems worldwide. This is known as an uncompressed digital voice signal.

Uncompressed digital voice would require as much frequency bandwidth to transmit as the FM voice signal used in analog mobile wireless phone systems. However, using digital voice compression technology, known as vocoders, the required data rate can be reduced. There is a trade-off in compressed bit rate, voice quality, and complexity of voice compression algorithm used. In North American TDMA systems, the voice was generally compressed from 64 to 8 kbps, using an algorithm known as vector sum excited linear prediction (VSELP). Unfortunately, the voice quality of VSELP was poorer than the previous analog FM-modulated AMPS system. Subsequently, using more powerful DSPs both in base stations and mobile handsets, a more powerful 8 kbps voice compression algorithm known as advanced code excited linear prediction (ACELP) was used, which closed the quality gap. Both vocoders used convolutional encoding and Viterbi decoding error correction, which resulted in a transmitted data rate of about 13 kbps.

13.5 TDMA Modulation and Demodulation

In the United States, the TDMA upgrade system to the AMPS system was known as IS-54 and then later upgraded to IS-136, referring to the Interim Standards of the telecommunications industry association (TIA). This system used the same 30 kHz channel spacing as AMPS. Each frequency channel was organized into frames of 20 ms, or 50 frames per second. Each frame had three time slots, each of which can be assigned to one user, as shown in Figure 13.2. This capability increased the capacity of the system threefold, compared to AMPS. Since it was compatible and inclusive with AMPS, the digital service could be gradually rolled out by the service provider, allowing for a gradual obsolescence of the AMPS handsets.

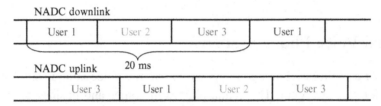

Figure 13.2

Notice that the frame timing is offset between the downlink and uplink. This offset allows the mobile handset to operate in transmit and receive modes at different, nonoverlapping intervals. Since the mobile handset needs to transmit for only about one third of the time, power consumption can be reduced and battery life extended.

A number of DSP technologies were used in TDMA systems. In addition to vocoding and error correction, baseband digital modulation and demodulation methods were implemented.

The channel quality between the base station and mobile handsets is often of very poor quality. Unlike with satellite and microwave links, with digital links it is rare to have a direct line of sight connection between base stations and handsets. The received signals are composed of multiple reflections, often distorted from passing through walls or other obstructions. These signals can sometimes combine out of phase, effectively canceling each other. This phenomenon is known as Rayleigh fading, which can be mitigated by using a second receive antenna (diversity). If the antennas are sufficiently separated in distance, then the phases of the multiple signals will vary differently, and in fact the Rayleigh fading will be uncorrelated, or independent. This means that the likelihood of the signal at one antenna canceling due to Rayleigh fading at the same time as the other receive antenna is very small. Through dynamic switching between the antennas, depending on which has the best signal, the impact of this fading can be largely mitigated, compared to using just a single antenna.

A reflection can become delayed and be received on top of other later symbols at the receiver. This is called inter-symbol interference (ISI) and can be compensated for by using an adaptive equalizer. Then there is the effect of the handset motion, which causes Doppler frequency shift in the received signals, and rapid changes in the ISI and fading effects, requiring fast-adapting digital receivers.

In IS-136, a form of QPSK modulation was used, called differential $\pi/4$ offset QPSK. It is more robust in terms of synchronization and sensitivity to Doppler shift compared to standard QPSK. It also has reduced dynamic range, which is beneficial to RF power amplifier performance. The receivers in both the base stations and handsets were equipped with adaptive equalizers, multiple or diversity receivers, synchronization, and frequency offset compensation algorithms. All this was implemented in DSP software in both the handsets and base stations.

There were alternative forms of TDMA. In Japan, a system called Personal Digital Communications (PDC) was widely used. It had 25 kHz channel spacing and also used three voice slots per 20 ms frame. Later, a half-rate vocoder with sufficient quality, called PSI-CELP, was developed; it allowed six voice slots per frequency channel. This was the most spectrally efficient version of TDMA. In Europe, the Global System for Mobile Communications (GSM) TDMA system was developed, which had eight voice slots, but used a wider 200 kHz channel spacing. While not especially spectrally efficient, it was adopted across all the European countries and came to have widespread commercial adoption around the world. Due to the simplicity and low cost of GSM handsets and infrastructure, it is still popular in the developing world, where often there is little or no landline phone service. The GSM system used Gaussian Minimum Shift Keying (GMSK) modulation, which is a form of phase shift modulation. Unlike QAM and QPSK modulation, GMSK is a constant amplitude type of modulation. Similar to FM, this made GSM signals very efficient to amplify for transmission in both the base stations and handsets.

The TDMA architectures, both handset and base station, were DSP based. The main difference was that the base station radios supported all the time slots simultaneously. Mobile handsets, on the other hand, would support one uplink and downlink time slot simultaneously, sufficient for a single call. Later, a derivative system called Enhanced Data GSM Evolution (EDGE) was developed from GSM. Sometimes called a "2.5G" technology, EDGE was developed to be an add-on to GSM networks; it supports somewhat higher data rates for mobile Internet or email access. It provides these higher date rates by allowing a single handset to occupy all 8 time slots of the frequency channel and by using a more efficient modulation method called 8-PSK.

Figure 13.3 shows a basic block diagram of a TDMA base station radio system. There are typically 4-24 radio cards in a TDMA base station.

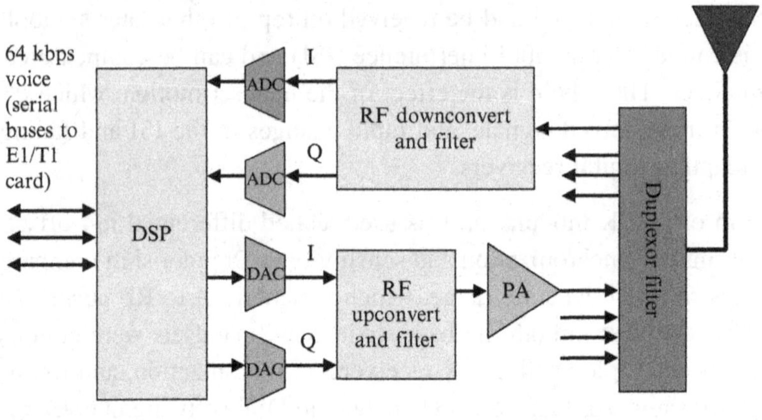

Figure 13.3

TDMA mobile phone networks work well, are relatively simple and low cost, and provide reliable wireless connectivity rates sufficient for compressed voice. However, new technologies promised higher data rates for Internet access, plus more efficient, higher-capacity networks. The latter became more important as wireless usage grew, and there was increased demand for frequency spectrum to service the market demand.

However, a rival technology was also being developed, known as Code Division Multiple Access (CDMA). The original goal of second-generation digital systems was a tenfold increase in capacity compared to AMPS. The IS-136 TDMA standard, with a threefold increase, fell short of that goal. Proponents of CDMA claimed that their technology would be able to meet this capacity increase. And that is the subject of the next chapter.

CHAPTER 14

CDMA Wireless Communications

The term *CDMA* stands for Code Division Multiple Access. CDMA modulation and demodulation technology grew out of military spread spectrum techniques. CDMA technology was commercialized by Qualcomm in the early 1990s. The initial Qualcomm CDMA system was known as IS-95. Later, Qualcomm developed and deployed an enhanced version of IS-95, known as CDMA2000 1xRTT, or just CDMA2000.

14.1 Spread Spectrum Technology

The basic idea behind spread spectrum is to start with a narrow band signal and mix it with a high-frequency pseudo-random number (PN) signal, which would have the effect of "spreading" the frequency spectrum occupied by the signal, thereby making it much harder for adversaries to jam or interfere (see Figure 14.1). Upon reception, the process could be reversed using the same pseudo-random code, and the original signal recovered.

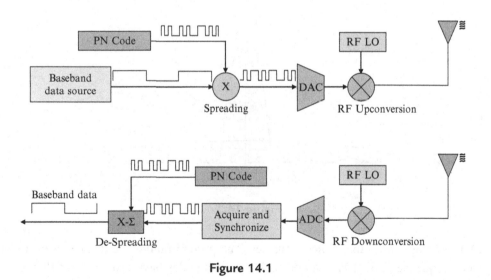

Figure 14.1

Digital Signal Processing 101. DOI: 10.1016/B978-1-85617-921-8.00018-3

14.2 Direct Sequence Spread Spectrum

There are several methods to perform frequency spreading. The method employed in CDMA is known as direct sequence spread spectrum. In direct sequence, the digital data is modulated by a much higher rate sequence of PN data. Each bit of the PN sequence is a "chip," and the higher rate is known as the chip rate. These chips typically modulate the much lower rate digital input data by a 180-degree phase shift in the carrier at the chip rate.

The chip rate phase changes are superimposed on the much lower rate phase shifts caused by input data (see Figure 14.2). The higher rate chip phase changes will greatly increase the occupied frequency bandwidth of the signal, and decrease the concentration of signal energy around the carrier.

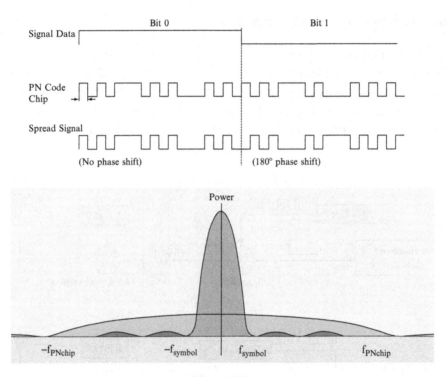

Figure 14.2

CDMA technology takes this concept further. The pseudo-random sequences were replaced by special sequences called Walsh codes. The Walsh codes have a property called orthogonalization. Each codeword is independent or orthogonal of the other. If any code is cross-correlated with another, the result is zero. This property is used to allow multiple users to share the same frequency band, each being assigned a unique codeword, which allows a receiver to pick out the one desired user signal from all the others (see Figure 14.3). The rest of other user signals are removed in the demodulation process.

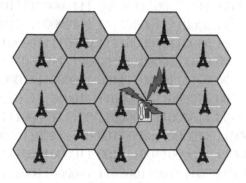

Figure 14.3

In CDMA mobile communications, not only do all the users in a given cell or sector share the same frequency channel, but all the cells also use the same frequency channel. For CDMA2000, each frequency channel is 1.25 MHz bandwidth. Unlike TDMA or analog systems, CDMA does not use frequency channel handoff as the user moves or transitions from sectors or cells. In fact, the mobile phone can be in communication with several cells simultaneously. This is one of the remarkable qualities of the CDMA mobile communications system.

14.3 Walsh Codes

There are 64 Walsh codes used in CDMA2000, each 64 bits long, listed here. The Walsh codes are clocked at the chip rate, which is 64 times faster than the data rate. In CDMA2000, the chip rate is 1.2288 MegaChips per second (Mcps), and the input data rate to the CDMA modulator is 19.2 kbps per Walsh code.

```
W0   0000000000000000 0000000000000000 0000000000000000 0000000000000000
W1   0000000000000000 0000000000000000 1111111111111111 1111111111111111
W2   0000000000000000 1111111111111111 1111111111111111 0000000000000000
W3   0000000000000000 1111111111111111 0000000000000000 1111111111111111
W4   0000000011111111 1111111100000000 0000000011111111 1111111100000000
W5   0000000011111111 1111111100000000 1111111100000000 0000000011111111
W6   0000000011111111 0000000011111111 1111111100000000 1111111100000000
W7   0000000011111111 0000000011111111 0000000011111111 0000000011111111
W8   0000111111110000 0000111111110000 0000111111110000 0000111111110000
W9   0000111111110000 0000111111110000 1111000000001111 1111000000001111
W10  0000111111110000 1111000000001111 1111000000001111 0000111111110000
```

W_{11} 0000111111110000 1111000000001111 0000111111110000 1111000000001111
W_{12} 0000111100001111 1111000011110000 0000111100001111 1111000011110000
W_{13} 0000111100001111 1111000011110000 1111000011110000 0000111100001111
W_{14} 0000111100001111 0000111100001111 1111000011110000 1111000011110000
W_{15} 0000111100001111 0000111100001111 0000111100001111 0000111100001111
W_{16} 0011110000111100 0011110000111100 0011110000111100 0011110000111100
W_{17} 0011110000111100 0011110000111100 1100001111000011 1100001111000011
W_{18} 0011110000111100 1100001111000011 1100001111000011 0011110000111100
W_{19} 0011110000111100 1100001111000011 0011110000111100 1100001111000011
W_{20} 0011110011000011 1100001100111100 0011110011000011 1100001100111100
W_{21} 0011110011000011 1100001100111100 1100001100111100 0011110011000011
W_{22} 0011110011000011 0011110011000011 1100001100111100 1100001100111100
W_{23} 0011110011000011 0011110011000011 0011110011000011 0011110011000011
W_{24} 0011001111001100 0011001111001100 0011001111001100 0011001111001100
W_{25} 0011001111001100 0011001111001100 1100110000110011 1100110000110011
W_{26} 0011001111001100 1100110000110011 1100110000110011 0011001111001100
W_{27} 0011001111001100 1100110000110011 0011001111001100 1100110000110011
W_{28} 0011001100110011 1100110011001100 0011001100110011 1100110011001100
W_{29} 0011001100110011 1100110011001100 1100110011001100 0011001100110011
W_{30} 0011001100110011 0011001100110011 1100110011001100 1100110011001100
W_{31} 0011001100110011 0011001100110011 0011001100110011 0011001100110011
W_{32} 0110011001100110 0110011001100110 0110011001100110 0110011001100110
W_{33} 0110011001100110 0110011001100110 1001100110011001 1001100110011001
W_{34} 0110011001100110 1001100110011001 1001100110011001 0110011001100110
W_{35} 0110011001100110 1001100110011001 0110011001100110 1001100110011001
W_{36} 0110011001100110 1001100110011001 0110011001100110 1001100110011001
W_{37} 0110011001100110 1001100110011001 1001100110011001 0110011001100110
W_{38} 0110011010011001 0110011010011001 1001100101100110 1001100101100110
W_{39} 0110011010011001 0110011010011001 0110011010011001 0110011010011001
W_{40} 0110100110010110 0110100110010110 0110100110010110 0110100110010110
W_{41} 0110100110010110 0110100110010110 1001011001101001 1001011001101001
W_{42} 0110100110010110 1001011001101001 1001011001101001 0110100110010110
W_{43} 0110100110010110 1001011001101001 0110100110010110 1001011001101001
W_{44} 0110100101101001 1001011010010110 0110100101101001 1001011010010110
W_{45} 0110100101101001 1001011010010110 1001011010010110 0110100101101001
W_{46} 0110100101101001 0110100101101001 1001011010010110 1001011010010110
W_{47} 0110100101101001 0110100101101001 0110100101101001 0110100101101001
W_{48} 0101101001011010 0101101001011010 0101101001011010 0101101001011010
W_{49} 0101101001011010 0101101001011010 1010010110100101 1010010110100101
W_{50} 0101101001011010 1010010110100101 1010010110100101 0101101001011010

W_{51} 0101101001011010 1010010110100101 0101101001011010 1010010110100101
W_{52} 0101101010100101 1010010101011010 0101101010100101 1010010101011010
W_{53} 0101101010100101 1010010101011010 1010010101011010 0101101010100101
W_{54} 0101101010100101 1010010101011010 0101101010100101 1010010101011010
W_{55} 0101101010100101 0101101010100101 0101101010100101 0101101010100101
W_{56} 0101010110101010 0101010110101010 0101010110101010 0101010110101010
W_{57} 0101010110101010 0101010110101010 1010101001010101 1010101001010101
W_{58} 0101010110101010 1010101001010101 1010101001010101 0101010110101010
W_{59} 0101010110101010 1010101001010101 0101010110101010 1010101001010101
W_{60} 0101010101010101 1010101010101010 0101010101010101 1010101010101010
W_{61} 0101010101010101 1010101010101010 1010101010101010 0101010101010101
W_{62} 0101010101010101 0101010101010101 1010101010101010 1010101010101010
W_{63} 0101010101010101 0101010101010101 0101010101010101 0101010101010101

Each user's input data stream of 19.2 kbps is modulated by a different Walsh code.

14.4 Concept of CDMA

A common nontechnical analogy to CDMA is the following. Imagine a round table, where there are multiple one-to-one conversations occurring between various pairs of people who are not adjacent to each other. Ordinarily, this would present a difficult situation, and it would be very difficult for anyone to communicate due to interference from all the other conversations. Now imagine if each pair of people spoke only one language, and for each pair, it was a different language. Now the conversations could proceed much more efficiently. Each pair would hear the other pairs' conversations as unintelligible noise, and their own conversation would stand out because each person could correlate what he heard against familiar words and speech of his own language. Another caveat is that this communication will work only if there are not too many other conversations and if everyone cooperates by speaking in a conversational tone at the same volume.

If one pair tries to enhance their conversation by raising their voices, it degrades everyone else's conversation. And if others in turn respond by raising their voices, things soon degenerate into a shouting match, and all communication is hindered. With this analogy in mind, we try to outline the essential basics of the CDMA2000 system in the following sections.

14.5 Walsh Code Demodulation

Imagine each Walsh code W_k sequence is mapped so that a zero is a $+1$, and a one is a -1. There is a single user channel input data bit every 64 chips, or period, of the Walsh code. If the data bit is "0," then the Walsh code W_k is transmitted, or if it's "1," the inverse, or negative, of the W_k is transmitted. Furthermore, the detector is based on a correlator, or integrator. The correlator performs a correlation (cross-multiply and sum) of the received sequence against the same Walsh code W_k. Next, let's go through a few examples. To make them easier to represent, we use only the first 16 chips of each Walsh code and pick the Walsh codes that are orthogonal over the first 16 chips. These are every fourth Walsh code, numbers 0, 4, 8, 12…60. This set of 16 sequences, each 16 chips long, forms a set of 16 orthogonal Walsh codes. The same concept applies to the larger set of 64 Walsh codes and when the correlation is applied over the complete 64 chips.

Example 1:

Input bit = 0, with Walsh code W_0, gives a transmitted signal of

+1, +1, +1, +1, +1, +1, +1, +1, +1, +1, +1, +1, +1, +1, +1, +1

When we correlate against shortened W_0,

```
   +1, +1, +1, +1, +1, +1, +1, +1, +1, +1, +1, +1, +1, +1, +1, +1
X  +1, +1, +1, +1, +1, +1, +1, +1, +1, +1, +1, +1, +1, +1, +1, +1
   ────────────────────────────────────────────────────────────
   +1, +1, +1, +1, +1, +1, +1, +1, +1, +1, +1, +1, +1, +1, +1, +1
```

$\Sigma = +16$, which we decode as a "0" input bit.

Example 2:

Input bit = 1, with Walsh code W_0, gives a transmitted signal of

−1, −1, −1, −1, −1, −1, −1, −1, −1, −1, −1, −1, −1, −1, −1, −1

When we correlate against shortened W_0,

```
   +1, +1, +1, +1, +1, +1, +1, +1, +1, +1, +1, +1, +1, +1, +1, +1
X  −1, −1, −1, −1, −1, −1, −1, −1, −1, −1, −1, −1, −1, −1, −1, −1
   ────────────────────────────────────────────────────────────
   −1, −1, −1, −1, −1, −1, −1, −1, −1, −1, −1, −1, −1, −1, −1, −1
```

$\Sigma = -16$, which we decode as a "1" input bit.

Example 3:

Input bit = 0, with Walsh code W_{32}, gives a transmitted signal of

+1, −1, −1, +1, +1, −1, −1, +1, +1, −1, −1, +1, +1, −1, −1, +1

When we correlate against shortened W_{32},

$$+1, -1, -1, +1, +1, -1, -1, +1, +1, -1, -1, +1, +1, -1, -1, +1$$
$$X +1, -1, -1, +1, +1, -1, -1, +1, +1, -1, -1, +1, +1, -1, -1, +1$$
$$\overline{+1, +1, +1, +1, +1, +1, +1, +1, +1, +1, +1, +1, +1, +1, +1, +1}$$

$\Sigma = +16$, which we decode as a "0" input bit.

Example 4:

Input bit $= 1$, with Walsh code W_{52}, gives a transmitted signal of

$$+1, -1, +1, -1, -1, +1, -1, +1, -1, +1, -1, +1, +1, -1, +1, -1$$

When we correlate against shortened W_{52},

$$+1, -1, +1, -1, -1, +1, -1, +1, -1, +1, -1, +1, +1, -1, +1, -1$$
$$X -1, +1, -1, +1, +1, -1, +1, -1, +1, -1, +1, -1, -1, +1, -1, +1$$
$$\overline{-1, -1, -1, -1, -1, -1, -1, -1, -1, -1, -1, -1, -1, -1, -1, -1}$$

$\Sigma = -16$, which we decode as a "1" input bit.

Next, let's assume we have three different codes in use—W_0, W_{32}, W_{52}—and have input bits 1 for W_0 (example 2), 0 for W_{32} (example 3), and 1 for W_{52} (example 4). Next, we sum the transmitted signals together, for a combined signal to be sent to the receiver:

$$-1, -1, -1, -1, -1, -1, -1, -1, -1, -1, -1, -1, -1, -1, -1, -1$$
$$+ +1, -1, -1, +1, +1, -1, -1, +1, +1, -1, -1, +1, +1, -1, -1, +1$$
$$+ +1, -1, +1, -1, -1, +1, -1, +1, -1, +1, -1, +1, +1, -1, +1, -1$$
$$\overline{+1, -3, -1, -1, -1, -1, -3, +1, -1, -1, -3, +1, +1, -3, -1, -1}$$

In the receiver, we can recover the original bits correlating to the original Walsh codes.

When we correlate with shortened W_0,

$$+1, +1, +1, +1, +1, +1, +1, +1, +1, +1, +1, +1, +1, +1, +1, +1$$
$$X +1, -3, -1, -1, -1, -1, -3, +1, -1, -1, -3, +1, +1, -3, -1, -1$$
$$\overline{+1, -3, -1, -1, -1, -1, -3, +1, -1, -1, -3, +1, +1, -3, -1, -1}$$

$\Sigma = -16$, which we decode as a "1" input bit.

When we correlate against shortened W_{32},

$$+1, -1, -1, +1, +1, -1, -1, +1, +1, -1, -1, +1, +1, -1, -1, +1$$
$$X +1, -3, -1, -1, -1, -1, -3, +1, -1, -1, -3, +1, +1, -3, -1, -1$$
$$\overline{ +1, +3, +1, -1, -1, +1, +3, +1, -1, +1, +3, +1, +1, +3, +1, -1}$$

$\Sigma = +16$, which we decode as a "0" input bit.

When we correlate against shortened W_{52},

$$+1, -1, +1, -1, -1, +1, -1, +1, -1, +1, -1, +1, +1, -1, +1, -1$$
$$X +1, -3, -1, -1, -1, -1, -3, +1, -1, -1, -3, +1, +1, -3, -1, -1$$
$$\overline{ +1, +3, -1, +1, +1, -1, +3, +1, +1, -1, +3, +1, +1, +3, -1, +1}$$

$\Sigma = +16$, which we decode as a "1" input bit.

This simple example shows how a composite of several Walsh codes can be separated into individual contributions by the correlation process, and the input bits used to set the polarity of the individual Walsh codes recovered. During this recovery process, the other Walsh codes are completely excluded due to the nature of orthogonality of the codes.

Next, let's consider what happens when we correlate the received sequence against a Walsh code that is not present in the composite signal.

When we correlate against shortened W_{20},

$$+1, +1, -1, -1, -1, -1, +1, +1, -1, -1, +1, +1, +1, +1, -1, -1$$
$$X +1, -3, -1, -1, -1, -1, -3, +1, -1, -1, -3, +1, +1, -3, -1, -1$$
$$\overline{ +1, -3, +1, +1, +1, +1, -3, +1, +1, +1, -3, +1, +1, -3, +1, +1}$$

$\Sigma = 0$. This indicates no correlation and no W_{20} component in the received signal.

The properties of Walsh codes allow each different code to be perfectly separated from the composite signal of all the Walsh-coded user data.

14.6 Network Synchronization

The concept described in the preceding section can be extended to build a receiver with a bank of correlators and threshold detectors to recover the data modulated by any given Walsh code. For the correlation process to work, all the Walsh codes must be transmitted with the same start and end timing. This process requires close synchronization between all the base stations and the mobile phones. The base stations are all synchronized using global positioning system (GPS) satellite receivers. In addition to position data, the GPS system also provides timing data with great precision. By comparison, the previous analog and TDMA cellular systems did not require base station synchronization. The mobile phones would reacquire each base station's individual timing at each handoff.

While use of the GPS system provides a means for all the CDMA base stations to be synchronized, this approach does not work for mobile phones (GPS signals cannot be received indoors or when blocked by tall buildings). Instead, a pilot signal is transmitted by each base station. The pilot signal provides a known signal to allow the mobile to determine the basic start and end time of the Walsh code timing, provides a coherent reference for demodulation, and allows for individual base station identification.

Recall in our discussion of Walsh codes that 180-degree phase shifts in the signal are used to identify $0\rightarrow1$ and $1\rightarrow0$ transitions. However, wireless signals can carry only relative phase information, not absolute (phase is the same as delay, so the absolute phase changes whenever distance between the transmitter and receiver changes). This limitation can be overcome in two ways. First is to encode all data in a differential form, such as the differential offset QPSK used in IS-136 TDMA. This is known as noncoherent demodulation. It is simpler but more susceptible to noise. The other method is to detect and track the received carrier phase and use this to determine actual phase changes. This is known as coherent demodulation. All this sounds complicated, and it is. Suffice to say here that the pilot signal is required to perform coherent demodulation and provides a phase reference for the correlators.

14.7 RAKE Receiver

The receiver architecture used for CDMA is known as a RAKE receiver. The RAKE receiver has multiple correlators. These correlators can be programmed for different Walsh codes. Some of the correlators can also be programmed for slight differences in arrival timing (remember multipath in the preceding chapter?) of the same Walsh code. The CDMA receiver constantly tunes the RAKE receiver multiple correlators, also called "fingers," for optimal reception. This is required in the presence of multiple delayed versions of the received signal or multipath, which would cause ISI in a TDMA system. Multipath is much more prevalent in CDMA, due to the much higher chip rate than TDMA's symbol rate, and the RAKE receiver architecture is well suited to compensate for this effect.

14.8 Pilot PN Codes

PN sequences are important in CDMA systems. One key characteristic of PN sequences is the autocorrelation property. When a PN sequence correlates to itself, the result is a perfect match if the sequences are perfectly time aligned, as with any other type of sequence. But if a PN sequence is correlated with an offset, even just one sample, the result is near zero correlation. This is also a property of a true random noise signal of infinite length.

One of the most important uses of PN sequences is in the pilot signal. The pilot signal is a pair of PN sequences that operate at the chip rate and repeat about every 26.66 milliseconds. The pilot signal is sent in quadrature, using QPSK. There is a separate sequence defined for I and Q, given by the following generator equations:

I PN sequence: $G_I = 1 + X^2 + X^6 + X^7 + X^8 + X^{10} + X^{15}$
Q PN sequence: $G_Q = 1 + X^3 + X^4 + X^5 + X^9 + X^{10} + X^{11} + X^{12} + X^{15}$

The PN sequences can be generated using a very simple circuit, as shown in Figure 14.4. The PN circuit output repeats every 32,767 bits, and an extra zero is inserted to bring this to the full 32,768-length pilot sequence. The shift register state has 2^{15}, or 32,768, possible values. The only invalid state is the all-zero state because this state is self-perpetual (that is why the extra zero is added separately to the output sequence). Any nonzero state can be used at startup, and this state determines where the repetitive 32,767 sequence begins, or the startup "phase" of the PN sequence. The pilot signal, with the zero inserted, has 32,768 possible phases.

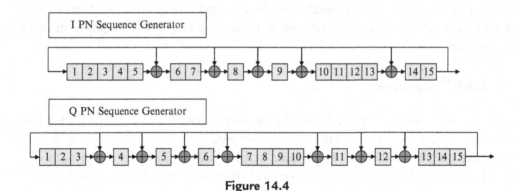

Figure 14.4

14.9 CDMA Transmit Architecture

The transmitter circuit for IS-95 and CDMA2000 is shown in the diagram in Figure 14.5. The I and Q quadrature paths are each mixed, or EXORed, with separate PN sequences. These PN sequences make up the pilot channel. Because the pilot channel uses W_0, it is the only channel not modified by the Walsh code mixing. The pilot tone is a simple QPSK signal, modulated by the pilot PN sequences. Different cells or base stations distinguish themselves by using different offsets, or phases, of the pilot sequence. The offsets are in increments of 512 chips, providing 64 possible offsets. When a CDMA mobile phone scans for the nearby cell sites, it identifies them by the relative signal strength and their PN offset of the pilot channel. Because transmission delays due to distance between base station and mobile are usually in the tens of chips, a separation in phase or delay of 512 chips clearly indicates that another base station is the source. These known PN sequence phase delays of 512 chips allow the mobile phone to perform a correlator search across increments of 512 chip delays to detect nearby base station pilots.

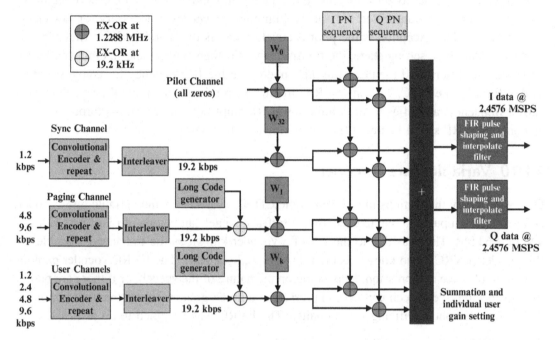

Figure 14.5

The mobile phones uses the strength of the pilot signal to determine if it is nearing the edges of the cell, defined by the area of strong pilot signal coverage. By measuring the strength of other phase offset pilots, the mobile phone can report to the network which of the adjacent cells are in range, information that is used in handoff decisions.

Note that the pilot I and Q sequences are used both in the pilot and all other channels, after Walsh code modulation. This provides for the individual quadrature components for each channel. The synchronization channel uses W_{32} and provides the GPS time reference used to obtain all the system timing information needed. Included in this is the data needed to set up the long code generator. This is a much longer PN sequence, 2^{42} bits long (which repeats about every 41 days!). Each mobile is assigned a unique long code phase to provide privacy for the user channel data. There is often only one paging channel, although the system allows up to seven paging channels. Paging channels are used similarly to control channels in

TDMA systems. The mobile phones monitor the paging channels to determine if there is an incoming call. The rest of the Walsh codes, up to 55, can be allocated to users for voice or data calls. Due to mutual interference limitations, the number of simultaneous users is normally 40 or fewer.

All the channels are then summed together. At this point, the signal is no longer a sequence of 1s and 0s, but is often 10–14 bits wide. In addition, gain is individually applied to each channel (or set to zero for unused channels). In CDMA, optimal capacity requires that each channel be transmitted only at the minimum power level necessary for low error rate reception. The exception is the pilot channel, which is discussed further. After the summation and gain setting stage, the transmit signal is then filtered and interpolated by the pulse-shaping filters, and then digitally upconverted to the IF frequency for conversion to analog form by the DAC. The analog RF circuitry then mixes and filters the signal to the carrier frequency, amplifies using a high-power RF amplifier, and transmits through the antennas. The RF signal bandwidth is approximately 1.25 MHz.

14.10 Variable Rate Vocoder

One of the many innovations introduced in the CDMA system was the variable rate vocoder. Notice that the input data rate to each user or traffic channel can be one of four rates: 1.2, 2.4, 4.8, or 9.6 kbps. This was due to the use of a vocoder known as the Enhanced Variable Rate Coder (EVRC), also known as the CTIA IS–127 standard. The EVRC vocoder exploits the nature of voice compression. Depending on the nature of the speech, or if there are pauses in conversation, the speech can be compressed to a much greater degree than the normal 8 kbps rate without sacrificing voice quality. The EVRC was designed to do just that.

With a TDMA system, the allocated bandwidth in a slot is fixed. A variable rate vocoder would need to be allocated a channel with a data rate equal to the maximum possible rate. CDMA, however, can take advantage of variable rates. When EVRC is operating at a rate of 1.2 kbps, the data can be repeated four times. This does not directly help the system operation. But with this repetition, the correlation is stronger because it is over four times the length. This, in turn, allows the data to be transmitted at lower power levels. This adjustment is possible because of the power, or gain, control adjustment for each individual channel. The very sophisticated power control algorithm keeps each mobile handset transmitting at the minimum power level to maintain a reasonable bit error rate. When EVRC is operating at a low data rate, the power level of that user channel can be reduced. Recall that in the uplink, each channel acts like noise in all the other channels. The lower the power required for each channel, the less noise-like interference is experienced by all the channels. For that reason, CDMA is often referred to as an interference limited system. The level of mutual interference, or noise, created by all the channels with respect to each other is the major determining factor in the capacity of a CDMA system.

14.11 Soft Handoff

The CDMA system uses a common 1.25 MHz frequency band in each cell site and sector. Through the use of different Walsh codes, interference is prevented between different users in the same cell or adjacent cells. For adjacent cells, even where the same Walsh code is in use, the different offsets of the short PN code limit interference by making the other users appear as noise. This use of the same frequency band allows "soft" handoff. In the preceding chapter, frequency handoff was discussed in analog and TDMA systems. This is called hard handoff, where the mobile phone must break communication with the existing base station and change frequency to communicate with the newly assigned base station. Soft handoff, in contrast, allows the mobile phone to maintain communication with the previous base station while simultaneously communicating with the next base station. Depending on the pilot strength thresholds set by the network, it is possible for a single mobile to be in communication with three or more base stations at once. The mobile phone has pilot signal strength thresholds to allow communication with a new base station and thresholds to discontinue communication with a previous base station. These thresholds are under network control.

Soft handoff can provide greater reliability and voice quality, due to the ability to communicate with both or several base stations at or near cell boundaries, where the signals tend to be weak. There is also a network capacity trade-off in that as the number of simultaneous base stations in the handoff process increases, or the duration of the soft handoff process increases, more network capacity is being consumed by that particular user.

Soft handoff is possible in CDMA networks because of the common frequency band used by all cells, and the network synchronization of all the base stations, neither of which exists in analog or TDMA systems. The network synchronization allows the voice or data to be simultaneously and synchronously routed to several base stations for transmission. The network can also synchronously combine voice or data traffic from a given mobile phone. This is an important advantage of CDMA voice systems over 2G systems.

14.12 Uplink Modulation

CDMA systems are so named for the downlink modulation method used. The uplink is somewhat different. The Walsh codes are orthogonal only if all the codes are sent and received synchronously. This is possible with the downlink because the base stations are all synchronized through GPS. In the uplink, the arrival time of the signals at the base station receiver are delayed by the roundtrip transit time. There is a time or propagation delay from the base station downlink signal to the mobile phone. It is the speed of light, which works out to about one mile per microsecond (millionth of a second), which is close to the one chip delay in CDMA2000. The mobile phone then synchronizes to the received signal using the downlink sync channel. The mobile, in turn, transmits the uplink signal, which also experiences a similar propagation delay to the base station. Because this is dependent on the mobile phone position, and constantly changing, it is not possible to guarantee that the different mobile phone uplink signals are all aligned at the base station receiver. Recall from the discussion on Walsh code demodulation, we needed all the 64-chip long Walsh code boundaries to be aligned to preserve the orthogonality. If one Walsh sequence is delayed by ½, 1, 2, or 5 chips, this demodulation process does not work. For this reason, CDMA modulation is not used in the uplink direction.

The uplink users are instead distinguished by different PN codes. Actually, they all use the same very long PN code but are assigned different phases. A 2^{42} length PN code is used, which is more than 4 trillion chips in length before repeating. Due to the autocorrelation properties of PN sequences, every other user uplink signal appears like noise to a receiver tuned to the correct PN sequence phase.

Walsh codes are still used in the uplink but not to separate the different users as in the downlink. In CDMA2000, there are 2^6, or 64, Walsh codes. In the uplink, after the convolutional encoder and interleaver, each set of 6 bits selects one of the 64 Walsh codes. The 64 chips of that Walsh code are mixed with the selected phase of the 2^{42} length PN code sequence. By detecting which of 64 Walsh codes was sent, the base station receiver can recover the original 6 bits. This process works similarly to maximum likelihood in coding. There is a possible 2^{64} sequences with a 64-bit word. However, only the 64 Walsh code sequences are valid transmit sequences. The base station receiver attempts to match the received sequence to the closest valid Walsh sequence.

After the Walsh encoding, the quadrature phases are mixed with I and Q PN short codes and pulse-shaped filters, similar to the downlink circuit. In this case, however, offset QPSK modulation is used rather than QPSK. The advantage of this approach is discussed in the next chapter.

The uplink is inherently the weaker link compared to the downlink. The Walsh codes in the downlink are perfectly synchronized and theoretically eliminate interference between users. But the uplink uses different PN phases to distinguish user channels. The autocorrelation properties do not eliminate each user's effect on each other, as the Walsh codes do, but make the other users' signals appear as noise. This makes the uplink interference limited. The downlink, on the other hand, tends to be power limited because the base station power amplifier must provide sufficient power to all users.

14.13 Power Control

Transmit RF power control was one of the key challenges of CDMA, and many of the enhancements of CDMA2000 over the earlier IS-95 system involved power control. Power control is used in both the downlink and uplink. In the downlink, the prime consideration is to keep the pilot power constant, even as the user channel power varies with the number of users and distance from the base station. Consistent pilot power is important because the pilot signal strength is used by mobile phones to determine cell boundaries and make handoff decisions.

Power control is even more critical in the uplink because each uplink is interference limited. To minimize the interference, each mobile phone needs to transmit an amount of power such that all the mobile transmit power levels are roughly equal *at* the base station antenna for equivalent user traffic bit rates (9.6 kbps).

Both open and closed loop power control methods are used. Open loop power control is performed by the mobile phone with no assistance from the base station network. Essentially, this type of control works as follows. The further from the base station, or more obstacles between, the lower the pilot and other downlink signal powers appear to the mobile phone. The mobile phone can use this information to estimate the uplink power level required. The weaker the pilot signal, the higher the loss and greater the distance to the base station, which requires a higher transmit power from the mobile to compensate. The converse is also true.

The rate of response of the open loop power control is nonlinear. If the received pilot signal suddenly increases, such as if the mobile emerges from behind a building, the mobile transmit power is immediately reduced. But if the pilot signal strength drops, then the mobile transmit power is increased slowly, to prevent inadvertent interference with other users' uplink signals.

In addition to open loop power control, there is also a closed loop power control loop operating. Closed loop power control is needed because estimating the uplink losses based on downlink losses can lead to errors. The downlink and uplink bands are different (separated by either 25 or 80 MHz depending on whether in the 800 or 1900 MHz bands), and different frequencies can fade independently.

Closed loop power control is more accurate because it is based on the uplink power measurements at the base station. A key challenge is to close the loop quickly. For that reason, a special power control bit is allocated in each downlink data frame, which can incrementally increase or decrease the mobile transmit power every 1.25ms. In addition, the base station is designed to use hardware circuits to drive this bit depending on receive correlation results.

Optimal uplink transmit power control requires quick reaction time be the basestation to tell the mobile how to adjust its transmit power level. In order to achieve rapid power control response, CDMA basestations use hardware circuits, rather than software-based control, to adjust the state of the power control bit.

14.14 Higher Data Rates

One advantage of CDMA is that multiple Walsh codes can be aggregated to a single user, to allow a higher data rate. Data rates in multiples of 9.6 kbps are available. Aggregating 16 channels together allows a data rate of 153.6 kbps. With 32 channels aggregated—the maximum because the pilot, sync, and paging channel are not available for user traffic—the data rate is 307.2 kbps. This rate is simply not possible in a TDMA system.

However, the disadvantage is that a few data users can consume most of the downlink capacity, compromising the available capacity for voice users, who provide the bulk of the revenue to the wireless carriers.

Therefore, a companion technology was developed to support high-speed data users. The market for this type of service is primarily businesspeople who need and are willing to pay for mobile high-speed service to allow remote Internet and email access. This service is known as Evolution for Data Only, or EVDO. It uses the same 1.25 MHz channel bandwidth and same uplink modulation techniques as CDMA2000. The downlink is completely redesigned and does not use CDMA techniques. It uses a very high speed form of QAM modulation, and the entire signal is devoted to one user at a time. The users share this high-speed link in a time-division duplex fashion. This link is suitable for high-speed data access, where the data access is often intermittent and packet based. The uplink typically requires much lower data rates and so retains the original CDMA uplink technology. Where EVDO service is offered, it uses a separate frequency spectrum from the CDMA2000 service.

14.15 Spectral Efficiency Considerations

So how does CDMA stack up against TDMA when it comes to spectral efficiency? The original goal was for digital systems to provide a 10-fold increase in spectral efficiency. The IS-136 TDMA efficiency provided a threefold increase.

- AMPS: One voice call per 30 kHz channel bandwidth. With seven-cell frequency reuse, the effective spectrum per user was 210 kHz.
- IS-136: Three voice calls per 30 kHz channel bandwidth. With seven-cell frequency reuse, the effective spectrum per user was 70 kHz.
- PDC: Six voice calls (half rate vocoder) per 25 kHz channel bandwidth. With seven-cell frequency reuse, the effective spectrum per user was 29 kHz.
- GSM: Eight voice calls per 200 kHz channel bandwidth. With seven-cell frequency reuse, the effective spectrum per user was 175 kHz.
- CDMA2000: CDMA is tougher to calculate capacity because this depends on network settings. In general, higher capacity is possible at the expense of voice quality, due to

the mutual interference issues and soft handoff thresholds. In general, most CDMA systems operate with a maximum of about 40 voice calls per cell. Assuming 40 voice calls per 1.25 MHz, with reuse every sector, the effective spectrum per user is 31 kHz.

To reach the 10-fold increase, the CDMA system would have to operate at the maximum of 55 voice channels per sector, which, while possible, does not lead to satisfactory quality in practice. But CDMA does lead to about a factor of over twice the spectral efficiency of IS-136 TDMA. The GSM TDMA system is by far the least spectrally efficient system. It now tends to dominate where lowest cost of service is critical, and user density (and therefore spectrum requirements) is moderate, such as in third world countries to provide basic phone service in rural areas.

14.16 Other CDMA Technologies

Alternate CDMA technologies were developed after CDMA2000. The most common CDMA system is known both as WCDMA, or Wideband CDMA, and UMTS. This form of CDMA was heavily based on IS-95 technologies. The biggest difference is a chip rate of 3.84 Mcps, a spectral bandwidth of 5 MHz, and a higher count of 256 Walsh codes to allow more user or traffic channels. This system was developed principally by the European wireless OEMs, and designed to be a 3G upgrade from the GSM system. UMTS has basically become the worldwide standard for CDMA because CDMA2000 is largely limited to usage in North America, Japan, and South Korea.

Just as Qualcomm developed EVDO for high-speed data access, a companion service to UMTS for high-speed data users was developed. It is known as HSDPA.

Another higher-rate version of CDMA2000 was developed by Qualcomm, called 3xRTT. This system had a chip rate of 3.68 Mcps, three times that of 1xRTT. It was designed to compete with the higher chip rate of WCDMA. It was never really deployed because similar capacity could be achieved using three separate 1xRTT systems.

A third CDMA system was developed in China. It is known as TD-SCDMA and is expected to be limited to deployment within China only.

CHAPTER 15

OFDMA Wireless Communications

The latest mobile communication technology following CDMA wireless systems is naturally called fourth generation, or "4G." The goals of a fourth generation system are yet even higher data capabilities, to be able to support voice, Internet, streaming video, and other services. Known as orthogonal frequency division multiplexing, or OFDM, it utilizes a completely different technology than 3G. OFDM has also been used in broadcast systems and for wireless LAN (Wi-Fi). These systems are point-to-multipoint systems. In the case of broadcast, it was basically one-way transmission. To use OFDM in mobile communications systems require a multiple access system, which requires some additional considerations. This is known as orthogonal frequency division multiple access, or OFDMA.

15.1 WiMax and LTE

There are two major standards for OFDMA mobile technology. The first is known as WiMax. It grew out of the Wi-Fi wireless LAN technology. Wi-Fi is now standard in virtually every PC laptop, DSL, or cable ISP gateway, providing private Internet access in many homes and public coverage in most airports, coffee houses, and hotels. Wi-Fi is defined as IEEE standard 802.11. WiMax, which has received major additions to support mobility and multiple access, is defined as IEEE standard 802.16. Shortly after the standardization of WiMax, the mobile communications industry began definition of its own OFDMA standard, which is known by the acronym LTE, for long term evolution. The technology path envisioned is a worldwide wireless technology roadmap for mobile service providers, starting with GSM, migrating to WCDMA, and eventually to LTE. Because LTE is being promoted by most of the mobile wireless industry, it is expected to have much wider deployment than WiMax. WiMax is more likely to find use in wireless network backhaul, military communications, and wireless local loop (basic phone service for rural areas without landline phones).

OFDMA does utilize a common concept with CDMA: orthogonality. To maximize spectral efficiency, OFDMA makes all the user traffic channels orthogonal to each other. In that way, there is no interference between users, even though they all share a large common frequency bandwidth. Yet in OFDMA, the different users do occupy different subsections of the frequency band, but are spaced much more closely together, compared to TDMA by using a technique that prevents adjacent channel interference.

Digital Signal Processing 101. DOI: 10.1016/B978-1-85617-921-8.00019-5

15.2 OFDMA Advantages

It is natural to ask what advantages OFDMA offers over CDMA mobile technology. First, OFDMA can be easily configured to support multiple bandwidths and, therefore, system capacities. LTE, for example, is able to operate in frequency bands ranging from 1.25 up to 20 MHz wide. CDMA, in contrast, has a fixed bandwidth, which is largely due to the chip rate and filtering. This characteristic allows OFDMA system operators to deploy service initially in a small frequency band and then expand that band as the number of users or customers increases. It also allows for higher data rates because more RF bandwidth can be used (up to 16 times CDMA2000 or four times WCDMA). Second, the individual subchannel modulation method can change dynamically depending on the quality of RF channel between the base station and mobile phone (which depends on the distance to base station, obstacles between, number of reflecting signals, and elevation and motion of user). OFDMA has the capability to vary on-the-fly the modulation used to either provide more robust communications links or a higher data rate on a per user basis. This is done without changing the amount of frequency bandwidth or number of subchannels that an individual user is allocated. Another consideration is more business focused. Given the prominent role Qualcomm had in the design, commercialization, and deployment of CDMA mobile technology, the company naturally enjoys dominance in the CDMA handset chip market, as well as the intellectual property rights and the associated royalties worldwide. OFDMA mobile technology, by comparison, is being more or less developed in parallel by multiple wireless and semiconductor companies and therefore provides a more level business landscape. Actual OFDMA system user capacity, spectral efficiency, and quality of service relative to CDMA are not known at the time of this writing because OFDMA widescale deployment is still in very early stages. Comparisons to OFDM used in the broadcast industry may not be relevant because this is a one-to-many broadcast system with little or no reverse-link traffic. In OFDMA mobile communications, the uplink is expected to be the weaker link, due to its multiple access requirement. However, if a significant portion of the mobile user traffic becomes Internet access or video streaming, then the downlink traffic loads will be much higher than the uplink, and the uplink limitations may be less important than downlink capacity.

In an FDD system, each user channel occupies a separate frequency band. Other users are rejected at the receiver by downconverting and filtering out the desired signal. Since all filters require a transition region between the passband and stopband, there must be a guard band, or separation region, between the frequency band of the individual carriers. The larger the difference in signal levels of multiple carriers at the receiver, the more rejection is required by the filtering. More rejection or attenuation of adjacent channels in the digital receive filters usually requires either a longer filter or a larger transition band.

With OFDM, we pack the different frequency channels, or subcarriers, very close together, and both demodulate all the subcarriers simultaneously. The basis of OFDM is utilization of the property that a group of sinusoidal signals spaced at a specific frequency separation will be orthogonal, or independent of each other. Just as we saw in CDMA, orthogonal signals do not interfere with each other, and using proper techniques, the desired signal can be separated from the other signals. We do not use traditional filtering to perform this separation of subcarriers.

15.3 Orthogonality of Periodic Signals

Two signals are considered to be orthogonal to each other when the cross-product of the two signals (e.g., sinusoids) equals zero. This is usually defined for periodic signals (signals which repeat) and the interval which the cross multiplication (or correlation) is the period of the signals.

What this means is that if we cross-multiply the two signals and add the results over a length of time in which both signals repeat, the result is zero. We saw this happen with the CDMA2000 Walsh codes in the preceding chapter. The period in that case was defined as 64 samples.

In continuous time, we can express the orthogonality relationship as

$$\int_0^T \cos(2\pi \, m \, f_0 t) \cdot \cos(2\pi \, n \, f_0 \, t) dt = 0, \text{ as long as } m \neq n$$

In this expression, m and n are integers, and $1/f_0 = T$, the period, which is the interval to integrate over.

We can see this scenario graphically in Figure 15.1. The fundamental frequency f_0 in the graph is 1 kHz, and the period T is 1000 μs or 1 ms.

Next, the product, or cross-correlation, of several cosines is graphed, as shown in Figure 15.2.

The results in Figure 15.2 are as follows:

    E: 1 kHz * 1 kHz, will integrate to a positive value over the interval T
    F: 1 kHz * 2 kHz, will integrate or average to zero over the interval T
    G: 1 kHz * 3 kHz, will integrate or average to zero over the interval T
    H: 1 kHz * 4 kHz, will integrate or average to zero over the interval T

Similarly, the 2 kHz sinusoid multiplied by the 3 kHz sinusoid, and every other combination of different frequency cosines, also integrates to zero over the integral T. Note that all the sinusoids are periodic over interval T, which simply means that they have an integer number

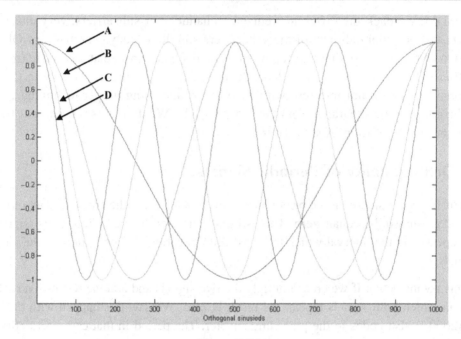

Figure 15.1
A: 1 kHz, B: 2 kHz, C: 3 kHz, D: 4 kHz

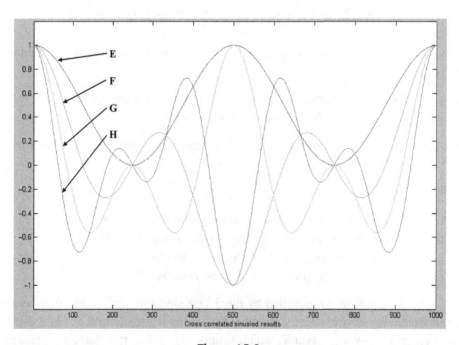

Figure 15.2

of cycles in that interval. Notice in Figure 15.1 that since every sinusoid returns to its starting value of "1" at the end of the 1000 μs interval, they are all periodic over T, equal to 1000 μs in this example.

In digital signal processing, we use sampled signals rather than continuous signals. The orthogonality relationship can be expressed as follows:

$$\sum_{k=0 \text{ to } N-1} \cos(2\pi \, m \, k/N) \cdot \cos(2\pi \, n \, k/N) = 0, \text{ as long as } m \neq n$$

For example, suppose we are sampling at 1 MHz. We can make the symbol period equal to 1000 samples, which equates to 1 k-symbol/second. Because of the orthogonality principle, we can have carriers at 1, 2 kHz, and so on, up to the Nyquist limit of 500 kHz. Each of these subcarriers can carry one symbol, perhaps using the phase of the subcarrier to carry the information. The demodulation process would require cross-multiplying, or cross-correlating, by each subcarrier frequency. This excludes the other subcarriers and allows the recovery of the symbol information by determining the phase of the subcarrier. In this way, it is possible to transmit simultaneous data across multiple subcarriers without interference.

15.4 Frequency Spectrum of Orthogonal Subcarrier

Each subcarrier is modulated, usually using either QPSK or QAM modulation. Sinusoids have their frequency content all at the carrier frequency and appear as a vertical line in the frequency spectrum. A QPSK or QAM modulation has nulls in the spectrum at the offset by the symbol frequency. In OFDM, these nulls line up perfectly with the adjacent subcarriers, resulting in minimal adjacent channel interference and efficient packing of the subcarriers in the frequency spectrum, as shown in Figure 15.3.

One possible method of OFDM modulation and demodulation would be to multiply each complex baseband QPSK or QAM signal by a complex exponential corresponding to each subcarrier frequency, located at integer spacings of the symbol frequency. For N carriers, this would require N parallel circuits for both OFDM modulation and demodulation.

Let us review the DFT and IDFT equations:

$$\text{DFT (time} \rightarrow \text{frequency)} \quad X_k = H(2\pi k/N) = \sum_{i=0 \text{ to } N-1} x_i e^{-j2\pi ki/N} \text{ for } k = \{0, \dots, N-1\}$$

$$\text{IDFT (frequency} \rightarrow \text{time)} \quad x_i = 1/N \cdot \sum_{k=0 \text{ to } N-1} X_k e^{+j2\pi ki/N} \text{ for } i = \{0, \dots, N-1\}$$

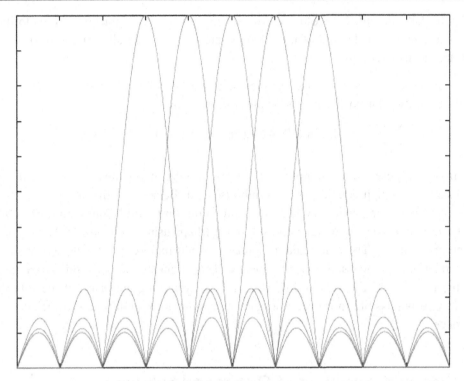

Figure 15.3

Expanding the IDFT equation, we get

$$x_0 = 1/N \cdot \sum_{k=0 \text{ to } N-1} X_k e^{+j0}$$

$$x_1 = 1/N \cdot \sum_{k=0 \text{ to } N-1} X_k e^{+j2\pi \cdot k/N}$$

$$x_2 = 1/N \cdot \sum_{k=0 \text{ to } N-1} X_k e^{+j2\pi \cdot 2k/N}$$

$$x_3 = 1/N \cdot \sum_{k=0 \text{ to } N-1} X_k e^{+j2\pi \cdot 3k/N}$$

$$\ldots$$

$$x_{N-1} = 1/N \cdot \sum_{k=0 \text{ to } N-1} X_k e^{+j2\pi(N-1)/N}$$

Notice that each complex exponential frequency is orthogonal to all the others. Each of these complex exponentials is periodic in N samples. The reason is that each has an integer number of cycles in N samples. To complete the orthogonality requirement, the symbol data rate should be equal to the frequency spacing, which is equivalent to the subcarrier spacing $e^{j2\pi \cdot 1/N}$, and will be $F_{sampling}/N$ symbols per second.

15.5 OFDM Modulation

The DFT can be used to perform OFDM modulation and demodulation. If we make N equal to 2^m, where y is an integer, then an even better way is to use the FFT. For example, if $m = 10$, then $N = 1024$. In OFDM systems, the IFFT is used for modulation in the transmit path, while the FFT is used for demodulation in the receiver. The OFDM modulation architecture is shown in Figure 15.4 for a 10 MHz LTE system.

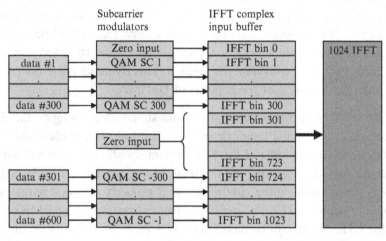

Figure 15.4

The IFFT processes 1024 complex samples to form a single OFDM symbol. Each user is assigned to a separate subcarrier. For 10 MHz LTE, there about 600 allowable subcarriers, spaced 15 kHz apart. Each subcarrier has a QPSK or QAM modulator that takes each user's input data and creates a single constellation point. This forms one bin of the IFFT input buffer. Each bin is mapped to an individual subcarrier in the frequency spectrum. Each OFDM symbol contains a single constellation point from each of the subcarrier modulators. And each successive OFDM symbol contains the successive constellation points from the modulators.

As shown earlier in the Figure 15.4, there is a data source for each subcarrier. This data source is modulated. If the chosen modulation for that subcarrier is QPSK, then every two bits of input data are mapped to one of four possible complex points on the QPSK constellation. The QPSK symbol for our subcarrier becomes one of the complex input samples for the IFFT. In parallel, this occurs with all the other subcarriers. They form the rest of the input samples for the IFFT. Additionally, certain samples in the IFFT input are always forced to zero. The IFFT output forms the OFDMA symbol. Therefore, the symbol rate of the

OFDMA system is the same as the rate at which the IFFT is performed. Selection of the input bin of the IFFT input buffer maps to a specific frequency subcarrier in the IFFT output. A continuous transmit data signal can be formed by concatenating successive IFFT outputs.

The process just described is the equivalent function as QSPK or QAM modulation of hundreds of users data in parallel and upconverting each using a different complex exponential with a frequency separation or difference equal to the QPSK or QAM symbol rate.

Note that each subcarrier can be independently modulated using either QPSK, 16-QAM, or 64-QAM. The higher order modulation requires a higher signal-to-noise ratio at the receiver, which occurs when there is little degradation in signal path from transmitter to receiver. The benefit is that higher amounts of data can be transmitted per subcarrier. Recall from Chapter 9 on modulation that QPSK carries 2 bits, 16-QAM carries 4 bits, and 64-QAM carries 6 bits of user data per symbol. This dynamic trade-off per user of data rate-to-signal link quality is one of the benefits of OFDMA.

The output of the IFFT forms the wide-band transmit signal. Depending on the IFFT size, the frequency spectrum can be made larger or smaller. The following table shows the configurable frequency bandwidths of the LTE system. In all cases, the subcarrier symbol rate and subcarrier frequency spacing are constant. Using a larger IFFT/FFT size allows for more subcarriers and occupies more bandwidth. This property allows a wireless service provider to initially deploy LTE in a smaller frequency spectrum, say 5 MHz. As customer usage grows, the LTE system can be reconfigured for up to 20 MHz. The mobile phones are dynamically configurable as well. A single subcarrier has bandwidth that is high enough to accommodate compressed voice data rates. However, higher data rates are available by assigning multiple subcarriers to the same user.

Table 15.1: LTE bandwidth dependent system parameters

BW (MHz)	1.4	3	5	10	15	20
Frame/slot length	10 ms / 500 µs	→				
Subcarrier spacing	15 kHz	→				
FFT size used	128	256	512	1024	1536	2048
Number of subcarriers	72	180	300	600	900	1200
Sampling frequency (MHz)	1.92	3.84	7.68	15.36	23.04	30.72
OFDMA symbol samples (using extended CP)	128 +32	256 +64	512 +128	1024 +256	1535 +384	2048 +512

Of course, an OFDMA system is a bit more complicated than this. For example, not all subcarriers are available for users. Certain subcarriers carry known, fixed data patterns and are known as pilot subcarriers. The receiver uses these subcarriers to perform synchronization, frequency offset tracking, and equalization.

The baseband signal has both positive and negative frequency subcarriers (remember, this is possible because we are using complex exponential subcarriers). The middle subcarrier, which is also the first IFFT bin, is located at 0 Hz in the baseband signal. It is always set to zero, to avoid introducing any DC into the baseband signal. And the outer subcarriers on both sides of the baseband signal are also set to zero, to provide a transition band for the low-pass filtering of the entire OFDMA signal from the last active subcarrier to the frequency channel edge (see Figure 15.5).

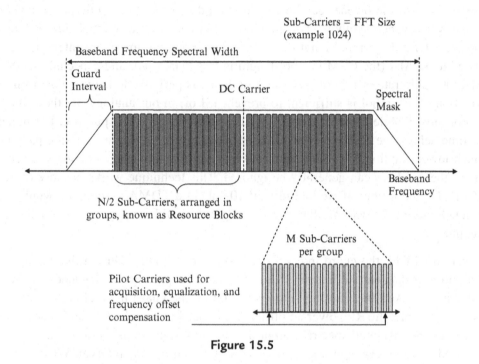

Figure 15.5

For example, the baseband sampling frequency of a 10 MHz LTE signal is 15.36 complex MSPS. So the Nyquist frequency is one half of this, or 7.68 MHz. The actual signal bandwidth is approximately $\pm 300 \cdot 15$ kHz $= \pm 4.5$ MHz. This frequency provides sufficient transition band for a low-pass filter (remember, there is another aliased image centered at each multiple of 15.36 MHz).

15.6 Intersymbol Interference and the Cyclic Prefix

Just as in the other wireless technologies, in mobile communications a direct line of sight is often impossible. This results in multiple reflected versions of the signal and of different signal strength, phase, and delay being received. This situation is sometimes referred to as "delay spread." The delay between different received reflections causes intersymbol interference, or ISI. In TDMA, adaptive equalizers are used to cope with ISI, and in CDMA, the RAKE receiver is designed to operate in the presence of ISI. OFDM systems use a different method, called a "cyclic prefix."

Of the three methods mentioned here, cyclic prefix is the simplest. The easiest way to prevent ISI is to simply pause between symbols. Think about trying to talk to someone at the other end of a long cave. If you yell, the echoes garble what the other person hears. But if you speak each word, separated by a pause of a few seconds, the echoes die away before the next word, and the listener can hear each word distinctly. The cyclic prefix involves a guard interval between symbols. This method works because the symbols are of long duration (for LTE the symbol length is about 67 μs, not counting guard interval), compared to TDMA (the GSM symbol length is less than 4 μs), and especially CDMA (CDMA2000 chip rate is 1.2288 MHz, or less than 1us period). So having a guard interval of a fraction of a symbol is sufficient to prevent ISI or, in our analogy, let the echoes die away. For most mobile environments, 5–10 μs is sufficient for this purpose. For much longer time delays, the received signal echoes take a much longer transmission path and so are much lower amplitude (perhaps reflected from a mountain range outside your city); therefore, these echoes can generally be ignored. This technique works because of the long OFDM symbol time. If we had to wait 10+ μs in a TDMA system, we would spend much more time waiting than transmitting and would have a very slow data throughput.

Still, even in OFDM, the guard interval does come at a penalty. During the guard interval, no user data can be transmitted, reducing the capacity and efficiency of the OFDMA system. And transmitting zero signal during the guard interval causes other issues. For example, the RF power amplifier would need to switch on and off each symbol, which can cause unwanted spectral output during these transitions. And this interval can still be used by the receiver if a known signal is sent during this interval. What is done in practice is that the last portion of the OFDM symbol is copied and inserted in the guard interval just prior to that OFDM symbol. This scenario is shown in Figure 15.6.

The receiver processes the OFDM symbol period including the guard interval. In LTE, for example, there is an extended cyclic prefix mode where the guard interval is set at 25% (the default mode is similar, but slightly more complex because the guard interval varies in different symbols). In this mode, the IFFT symbol period is 66.67 μs, with a guard interval of 16.67 μs, for a final symbol period of 83.33 μs. The receiver processes the OFDM symbol period of 83 μs including guard interval. During this interval, the symbol appears to be periodic or repeating. For this reason, the guard interval is called a cyclic prefix. Multiple OFDM symbols are aggregated to form slots and frames, which provide the structure needed to organize pilot subcarriers (used by mobile phones to synchronize timing and frequency to base station) as well as the user or data subcarriers.

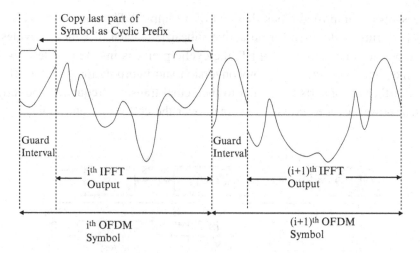

Figure 15.6

Figure 15.7 shows the organization of the LTE structure of frames, slots, and symbols, and how the samples are organized to allow for the cyclic prefix. This is shown for the 10 MHz bandwidth and extended cyclic prefix configuration. LTE systems are generally configurable to support all the different possible bandwidths and cyclic prefix options because this enables the wireless service provider to select the optimum configuration for its capacity and licensed spectrum allowances. The mobile phones are also able to support the various configurations.

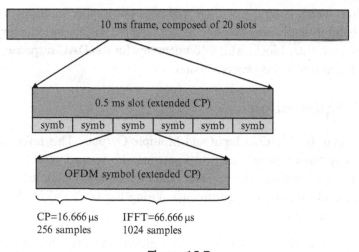

Figure 15.7

Figure 15.8 shows a simplified block diagram of a sample LTE transmitter. A bank of modulators performs modulation for all active subcarriers, each of which provides data to a different bin on the IFFT. After the IFFT, the cyclic prefix is inserted as described earlier. The rest of the chain performs digital upconversion and interpolation. Analog RF circuitry (not shown) further upconverts the signal to the actual transmit frequency and amplifies it to a sufficient power level to provide coverage over the cell sector and radius.

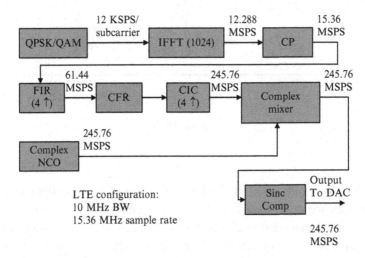

Figure 15.8

Two blocks in the diagram may be unfamiliar. The CFR block denotes crest factor reduction, which is discussed in the next section. The CIC block denotes a cascade integrate comb interpolation filter, which is a type of filter that does not require multipliers and so is inexpensive to build in hardware. The CIC is not covered in this introductory book. Note the sinc compensation filter block, which compensates for the DAC response, as described in Chapter 11 on digital up- and downconversion.

15.7 MIMO Equalization

MIMO is an acronym for Multiple Input and Multiple Outputs. This refers to antennas. Previous wireless systems typically used two antennas for reception and one antenna for transmission at the base station. The receiver could dynamically pick between the two antenna signals, greatly improving performance during Rayleigh fading. In MIMO systems, this approach is taken further.

For 4G wireless systems, the base stations have a minimum of two receive and two transmit antennas per sector. (Note that sometimes a transmit antenna and receive antenna are packaged together, particularly in FDD systems.) This is often referred to as 2×2 MIMO. Other likely configurations are 2×4 (two transmit and four receive antennas) and 4×4 (four transmit and four receive antennas).

In a MIMO system, the receiver does not simply select the best signal from two or more antennas. Instead, the receiver uses both signals and tries to estimate and compensate for the degradation the signal experiences in the separate paths to each antenna. This process usually involves solving multiple equations simultaneously and the use of matrix inversion algorithms to obtain the individual channel degradation estimates and perform the compensation, or equalization, of each receiver path.

In the uplink, the mobile signal travels to multiple receive antennas, with each antenna having independent, or uncorrelated, path reflections, fading, and additive noise. In the downlink, the mobile has only one receive antenna. However, multiple base station transmit antennas can be used in a similar manner. Often, each transmit antenna transmits a slightly different version of the signal, using a technique called "space time encoding," which is known by the mobile receiver. Each version of the signal experiences uncorrelated path reflections, fading, and additive noise, which can be simultaneously processed to obtain the best estimate of the transmit signal.

These techniques involve sophisticated processing and statistical theory and are not suitable for this introductory discussion.

The use 4×4 system basically doubles the cost of the most expensive portion of the wireless base station compared to a 2×2 system but promises improved performance and system capacity. Whether this improvement is enough to justify the additional base station cost is still unknown at the time of this writing.

15.8 OFDMA System Considerations

Before we conclude this introduction to OFDMA, it may be worthwhile to summarize the relative merits and challenges of this fourth generation mobile wireless technology.

In addition to the two benefits of configurable RF bandwidth (BW) and dynamically variable user data rates depending on subcarrier SNR described earlier in the chapter, there are some further benefits to OFDM technology.

OFDM efficiently deals with multipath, or ISI, using the cyclic prefix. This is much less complex and cheaper to implement than the adaptive equalizer in TDMA and RAKE receiver in CDMA.

OFDM is fairly insensitive to narrowband interference because only a few subcarriers are affected. And given the dynamic monitoring of subcarrier SNR, this situation can be detected and users allocated to subcarriers where interference is not present.

OFDM is, however, very sensitive to frequency offset. Both mobile and base station receivers must compensate for Doppler shift prior to demodulation to preserve orthogonality of the subcarriers.

OFDM is also computationally efficient when considering the amount of circuitry and number of calculations required to support high data rates. Due to the high efficiency of FFT implementations, mobile phones are able to receive high data rates with less DSP processing than comparable rates would require in a CDMA system, and would not be possible at all in a TDMA system.

15.9 OFDMA Spectral Efficiency

Estimating spectral efficiency for OFDMA, particularly for voice capacity, is difficult because no large high-capacity systems are in operation at the time of this writing. Many parameters can affect efficiency and capacity. For example, through use of dynamic modulation modes, multiple antenna receive and transmit (MIMO), and packetized voice (voice over IP), the proportion of voice to data traffic may contribute to system optimization.

Similar to CDMA, OFDMA systems use the same channel frequency in all cells (see Figure 15.9). This could be anywhere from a 1.25 to 20 MHz bandwidth. Interference between users is avoided by assigning different subcarriers to each base station. Groups of subcarriers are preassigned specific pilot subcarriers, so a mobile phone is able to use known pilot subcarriers to acquire and synchronize to any assigned group of subcarriers. The frequency reuse of the subcarriers is not defined and may vary by wireless service provider. In TDMA and analog systems, the frequency reuse pattern

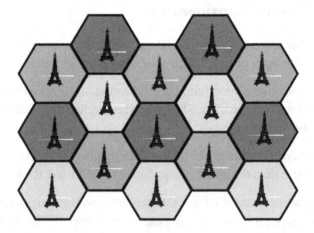

Figure 15.9

was every 7 cells. However, for OFDMA, it may be possible to have subcarrier frequency reuse of every 3 cells, due to the lower SNR requirements when using QPSK and MIMO techniques. Also, the channel spacing is tighter, a consequence of the OFDM modulation. For the LTE system, channel spacing is 15 kHz, much closer than any TDMA system.

In summary, at this point it is very difficult to reliably predict actual capacity and spectral efficiency with so many variables. Similar to CDMA, the OFDMA network service providers will have to go through a learning curve in optimizing their systems for both high capacity and high quality of service.

15.10 OFDMA Doppler Frequency Shift

All mobile communication systems must contend with Doppler frequency shift. However, OFDM is especially sensitive to Doppler shift because it relies on the precise alignment of subcarrier frequencies to provide orthogonality. The mobile phone velocity relative to the base station causes Doppler frequency shift, which must be tracked and compensated for to provide proper demodulation using FFT processing. The following equation can be used to calculate the Doppler frequency shift:

$$\text{Doppler Frequency shift} = f_{\text{carrier}} \cdot \text{relative velocity}/\text{speed of light}$$

For example, assume a system operating at a 2 GHz frequency band, with a mobile user traveling at a speed of 120 km/h (33.3 ms/s).

$$\text{Doppler Frequency shift} = 2 \cdot 10^9 \cdot 33.3/(3 \cdot 10^8) = 220\,\text{Hz}.$$

This speed is enough to cause a problem in OFDMA systems if not compensated for, by frequency shifting or rotating the input signal prior to demodulation.

15.11 Peak to Average Ratio

Both OFDM and CDMA have what is known as high peak to average ratios, or PARs, on the order 10–20 times, or 10–13 in decibels. Every signal has an average power. The peak to average power level, usually expressed in decibels, is the power level of the highest instantaneous power compared to the average power level. A PAR of 1, or 0 dB, means the signal is of constant power, so the peak power is equal to the average power. A large PAR means that the signal power fluctuated occasionally to a very large value; therefore, a large PAR requires the linear transmit amplification circuit to operate over a wide power range, which tends to be both costly and inefficient. The plot in Figure 15.10 shows a typical OFDM symbol amplitude over the duration of the symbol, with I and Q shown separately. This is a plot of the output of the IFFT in the modulator.

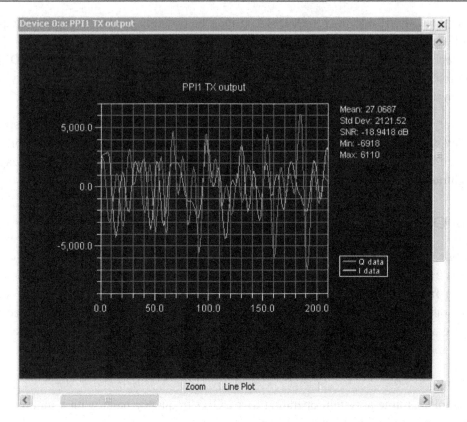

Figure 15.10

In mobile communication systems, the base station signals are amplified using fairly high-power amplifiers (PAs). The amplifiers must behave linearly over the output power range to avoid creating spectral energy outside the transmit band. The required power capacity and, therefore, power consumption and cost of the amplifier depend on the PAR of the signal being amplified.

Let us assume that we need to transmit 50 W of power, on average, to provide RF coverage to a cell during high-traffic hours. If the transmit signal has a PAR of 0 dB, we need an amplifier with a power rating of 50 W (able to output 50 W while operating linearly). However, if the signal has PAR of 3 dB, we ideally need an amplifier capable of 100 W to linearly amplify peaks of the signal. And if the signal has a PAR of 10 dB, we require an amplifier capable of a 500 W amplifier to linearly amplify the peaks of the signal. The amount of reduction in the input power to accommodate the peaks in the signal and still have linear performance in the PA is known in the industry as "backoff" and is expressed in decibels. If no compensation techniques are used, the required PA backoff is approximately the same as the input signal PAR. This assumes use of a basic class AB power amplifier.

The PAR directly effects PA efficiency. The efficiency is defined as the RF power output divided by the DC power input. High PAR tends to decrease efficiency because PA power consumption is roughly proportional to the peak RF output power capacity. High RF power output requires a high level of bias current at all times, which is reflected in the DC power consumption, even when the PA is not outputting high levels of output power.

Table 15.2: Power amplifier efficiency variance by technology

Wireless Technology	Wireless Standard	PAR (dB)	Typical PA Efficiency (%)
TDMA	GSM	0	60
TDMA	NADC	3	20
Multicarrier	any	10+	10*
CDMA	CDMA2000 or UMTS	10–12	10*
OFDM	LTE or WiMax	10–13	10*

*To achieve required linearity and meet adjacent channel spectral emission requirements, PAs often include "feed-forward compensation," which involves using analog techniques to cancel PA output distortion, improving linearity.

Note that GSM is 0 dB PAR. The reason is that the modulation technique used affects only the phase; there is no change in amplitude. This is one factor that makes GSM the lowest-cost wireless system and why it is still commonly used in many cost-sensitive markets, despite its low spectral efficiency.

15.12 Crest Factor Reduction

New digital technology now allows for improvements in PA efficiency. The two technologies are commonly known as crest factor reduction (CFR) and digital predistortion (DPD). A well-designed PA with CFR and DPD can achieve efficiency of about 30% in a typical OFDM application. This is a threefold increase in output power for the same PA circuit and power consumption. It results in major cost reductions and, due to lower power dissipation, higher reliability.

Crest factor reduction is a digital processing function that can reduce the PAR of a signal. A simple way to accomplish this is to limit or saturate the peaks in the digital signal amplitude. The problem with this approach is that it produces high-frequency spectrum components that cause interference in the adjacent frequency spectrum. More sophisticated techniques are needed. There are usually three different conditions to satisfy, with measurements associated with each:

- Maximum reduction in PAR of signal. PAR is measured using the complementary cumulative distribution function (CCDF) function on the spectrum analyzer.
- Minimum distortion of transmitted signal. Signal quality is measured using the EVM measurement mode of the spectrum analyzer.
- Minimum adjacent channel frequency signals. This is measured using an emission mask function on the spectrum analyzer.

Spectrum analyzers designed for wireless system development come with personality modules, which are firmware-enabled functions to perform many different measurements required by a specific wireless standard, such as LTE.

The complementary cumulative distribution function (CCDF) is used to measure PAR. A typical plot is shown in Figure 15.11.

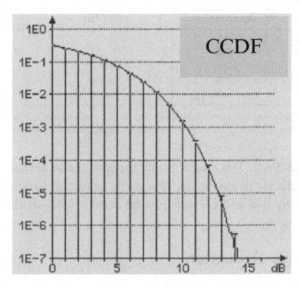

Figure 15.11

PAR is naturally statistical in nature. This is a plot of the percentage of time the signal power is a given number of decibels above the average power. For example, the plot here shows about 10% (1E-1) of the time the signal is 4.5 dB higher than average power. The plot also shows that about 1% (1E-2) of the time the signal is 8 dB higher than average power. The PAR is not just one value but varies according to the duty cycle of the time the signal is a given level above the average power. The maximum PAR depends on how long the measurement system can wait because the longer the measurement interval, the greater likelihood of a large peak in the signal to occur. Typically, for many wireless systems, the PAR is taken at about 0.001–0.0001%, or E-5 to E-6. In this case, the signal would have a PAR of about 13 dB. At percentages lower than this, higher peaks occur so infrequently to cause little effect in the other measurements, such as spectral mask or EVM measurements.

EVM stands for error vector measurement. A standard way to measure distortion in a signal is to measure the actual constellation points compared to the ideal constellation. In a perfect constellation, each point lands exactly on one of the allowable constellation values. CFR techniques introduce distortion. The EVM is defined as the distance between the actual demodulated point on the constellation and the nearest valid constellation, divided by the distance of the valid constellation point from the origin. Again, this is a statistical measurement, averaged over many points in a signal. It is normally expressed in percent. For example, for 64-QAM, the typical transmit quality must be 3% or less.

To prevent interference in the nearby spectrum, all wireless standards have a transmit emission mask (the GSM example in Figure 15.12) that specifies how quickly the transmit power must fall off as a function of the distance from the center of the RF carrier frequency. The measurement is usually specified in decibels, relative to the signal power at the carrier frequency. For example, in the GSM transmit mask requirement shown here, the RF power must drop by at least 30 dB at a 200 kHz offset to the carrier frequency.

Two common approaches are used in CFR, both applied to the digital baseband representation of the signal prior to upconversion. One is a time-domain-based technique,

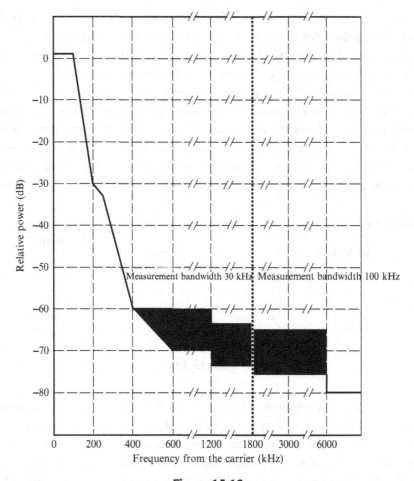

Figure 15.12

where peaks in amplitude are monitored and reduced in amplitude in a smooth manner to present unwanted high-frequency content from being added to the signal. This approach has the advantage of being able to be applied to any signal.

Another approach is often used in OFDMA systems such as WiMax or LTE. In this approach, the signal is converted to the frequency domain, and the amplitude of various subcarriers is adjusted in such a way as to reduce the CFR of the complete composite signal. After this adjustment, the signal is reconverted back to the time domain. For both of these techniques, the signal must first be interpolated to a sample rate of four or more times faster, which also proportionally enlarges the baseband frequency band. This is required because all CFR techniques introduce some higher frequency spectral content, which would be aliased into the baseband signal at a lower sample rate.

15.13 Digital Predistortion

A second technique used to increase PA efficiency is digital predistortion, or DPD. While CFR works to reduce the PAR of the input signal to the PA, the purpose of DPD is to extend the linear range of the PA, thereby reducing the amount of backoff required. Many proprietary techniques are used to compensate for the nonlinear behavior of a PA at high-output power levels. When a PA is overdriven or operated at output power levels beyond the linear range, it tends to saturate, which causes undesirable effects, such as generation of unwanted high-frequency distortion and degradation of EVM. PA saturation can be compensated for, by "predistorting" the input signal so that the PA output has the correct characteristics. This process requires a closed-loop operation, that is, feeding back an attenuated version of the PA output for monitoring by the DPD circuit (see Figure 15.13).

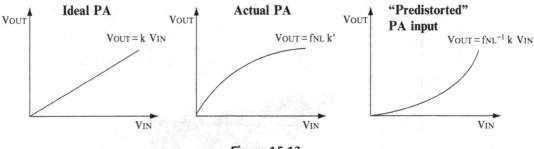

Figure 15.13

A very basic DPD circuit is depicted in Figure 15.14. It monitors the instantaneous power of the digital input signal and multiplies the input by polynomial with coefficient values stored in a register bank. The polynomial provides a non-linear output level at different input power levels. A closed-loop algorithm continually adjusts the register bank values to optimally compensate for the PA reduction in gain at high power levels, thereby predistorting the

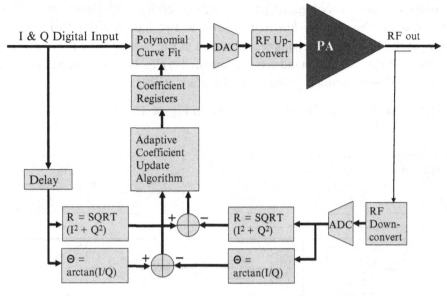

Figure 15.14

signal. Note that it cannot actually increase the RF power capacity of the amplifier but may increase the range over which the amplifier behaves linearly. One of the challenges of DPD is the multiple disciplines required. Digital signal processing, RF amplification, and software and hardware design knowledge are required.

Through use of DPD and CFR techniques, PA efficiency can be increased threefold. But there is yet another method to further increase the overall efficiency of wireless base stations that does not involve digital signal processing at all.

15.14 Remote Radio Head

For reasons for serviceability, cooling, and equipment size, a building or shed is used to house the base station at the base of an antenna tower. This setup typically requires RF cables of 50 m or longer between the base station and the antennas mounted on the antenna mast tower. This often results in a 2 dB or more reduction in transmit signal power, due to cable loss. This loss increases as the RF carrier frequency increases.

A 2 dB loss can mean almost two thirds of the RF amplifier power is dissipated in the cabling prior to reaching the antenna. To eliminate this loss, most modern wireless base stations are designed with remote radio heads (RRHs). The functions associated with the PA and antenna circuits are packed into an air-cooled module that is small and light enough to be mounted next to the antenna arrays, up to 50 m high in the antenna tower. This setup can reduce cost, primarily due to the use of much smaller PAs and elimination of RF cabling.

It does, however, require very high reliability because servicing RRHs mounted 50 m high is costly and difficult. The industry has standardized on two interface standards between RRH and the rest of the base station equipment mounted at the base of the tower or in a nearby building. These RRH interface standards are known as open base station architecture initiative (OBSAI) and common public radio interface (CPRI).

A simplified block diagram of an RRH is shown in Figure 15.15.

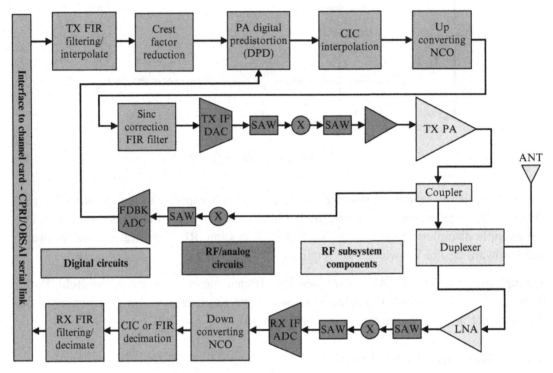

Figure 15.15

CHAPTER 16

Radar Basics

Radar stands for Radio Detection and Ranging. It can be used to detect target range, direction, and motion. Radar was first widely used during World War II and has continually evolved since. Early radars were composed of analog circuits, but since the 1960s, radars have become increasingly digital. Today, radar systems contain some of the most sophisticated and powerful digital signal processing systems anywhere.

Radar has widespread use in both commercial and military applications. Air traffic control, mapping of ground contours, and automotive traffic enforcement are just a few civilian applications. Radar is ubiquitous in military applications being used in air defense systems, aircraft, missiles, ships, tanks, helicopters, and so forth.

16.1 Radar Frequency Bands

Radar systems transmit electromagnetic, or radio, waves. Most objects reflect radio waves, which can be detected by the radar system. The frequency of the radio waves used depends on the radar application. Radar systems are often designated by the wavelength or frequency band in which they operate, using the band designations shown in the following table.

Table 16.1: Radar frequency bands

Radar Band	Frequency (GHz)	Wavelength (cm)
Millimeter	40–100	0.75–0.30
Ka	26.5–40	1.1–0.75
K	18–26.5	1.7–1.1
Ku	12.5–18	2.4–1.7
X	8–12.5	3.75–2.4
C	4–8	7.5–3.75
S	2–4	15–7.5
L	1–2	30–15
UHF	0.3–1	100–30

Digital Signal Processing 101. DOI: 10.1016/B978-1-85617-921-8.00020-1

The choice of frequency depends on the application requirements. The minimum antenna size is proportional to wavelength and inversely proportional to frequency.

Airborne applications often are limited in the size of antenna that can be used. A smaller antenna dictates a higher frequency and lower wavelength choice.

Beamwidth, or the ability of the radar to focus the radiated and received energy in a narrow region, is also dependent on both antenna size and frequency choice. Larger antennas allow the beam to be more tightly focused. Therefore, a higher frequency also allows the beam to be more tightly focused for a given antenna size. The "focusing" ability of the antenna is often described using an antenna lobe diagram, which plots the directional gain of an antenna over the azimuth (side to side) and elevation (up and down).

The range of the radar system is also influenced by the choice of frequency. Higher-frequency systems usually are lower power due to electronic circuit limitations, and they experience greater atmospheric attenuation. The ambient electrical noise that can impair operation of analog circuitry also becomes more pronounced at higher frequencies. Most of the radar signal absorption and scattering is due to oxygen and water vapor. Water vapor, in particular, has high absorption in the "K" band. When this was discovered, the band was divided into Ka, for "above" and Ku for "under," the frequencies where radar operation is limited due to water vapor absorption. At higher frequencies in portions of the millimeter band, oxygen causes similar attenuation through absorption and scattering.

Another consideration, discussed more fully in the next chapter, is the effect of the radar operating frequency on Doppler frequency measurements. Doppler frequency shifts are proportional to both the relative velocity and the radar frequency. Doppler frequency shifts can provide important information to the radar system.

Most airborne radars operate between the L and Ka bands, also known as the microwave region. Many short-range targeting radars, such as on a tank or helicopter, operate in the millimeter band. Many long-range ground-based radars operate at UHF or lower frequencies, due to the ability to use large antennas and minimal atmospheric attenuation and ambient noise. At even lower frequencies, the ionosphere can become reflective, allowing very long range over-the-horizon operation.

16.2 Radar Antennas

A critical function in any radar system is the antenna. Early radars often used mechanical parabolic antennas. The antenna is capable of focusing both the receive and transmit energy in a given direction (see Figure 16.1). The antenna could be moved mechanically using motors and aimed to search over different parts of the sky.

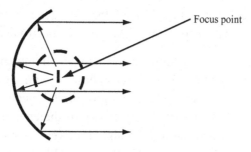

Figure 16.1

The degree of directionality is often shown in azimuth and elevation gain diagrams. The diagram in Figure 16.2 shows an antenna that has a fairly wide, or broad, main lobe. Most radar antennas may have a much narrower main lobe, on the order of a few degrees. Frequently, the width of the main lobe is specified by the point at which the receive or transmit signals are attenuated by 3 dB, or about one half. The antenna shown in this figure has a lobe width of about 20°. However, all antennas receive some level of signal from undesired directions, even from behind. The antenna gain plots visually quantify the relative gain across both azimuths and elevations (usually a separate plot for each). In general, the narrower the main lobe, the higher the antenna gain is.

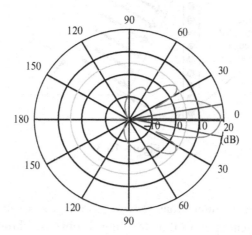

Figure 16.2

The antenna design influences both the amount of energy the radar can transmit at the desired target space, as well as how much energy it can receive from the same direction. It also determines how much unwanted energy from other directions is attenuated (e.g., reflections from the ground in an airborne search radar). Having a narrow or focused beam allows the energy to be more focused. To search across a wide area, the antenna must steer its beam across the entire search space. As just mentioned, this was done mechanically in early radars. However, more advanced radars, especially airborne, use electronically steerable antennas.

An electronically steerable antenna is built from many small antennas, or individual elements. Each element can individually vary the phase of both receive and transmitted signals, as well as the signal strength using analog or digital electronic circuits. It is the changes in phase that provide for steerable directivity of the antenna beam over both azimuth and elevation. Only when the receive signal arrives in-phase across all the antenna elements will the maximum signal be received. This provides the ability to "aim" the main lobe of the antenna in a desired direction. The process is reciprocal, meaning that the same antenna lobe pattern exists on both receive and transmit.

Each antenna element must have a delay, or phase adjustment, such that after this adjustment, all elements will have a common phase of the signal. If the angle $\theta = 0$, then all the elements receive the signal simultaneously, and no adjustment is necessary (see Figure 16.3). At a nonzero angle, each element has a delay to provide alignment of the wavefront across the antenna array.

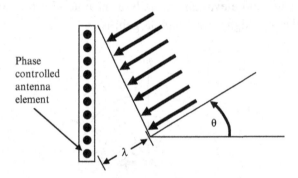

Figure 16.3

Using an electrically steered antenna has several advantages. It can be steered very rapidly, which can allow fast searching as well as tracking of objects. Through the use of a technique called "lobing," the radar beam can be rapidly steered on either side of a target. By noting where the stronger return is, the antenna can track the target location. Further, different regions of the antenna can be aimed in different directions to scan or track multiple regions or targets, albeit with a reduced transmit power and receive gain. A disadvantage of an electrically steered antenna is the reduced aperture at larger incident angles (see Figure 16.4). The aperture is one factor in the antenna gain, and decreases by the cosine θ, where θ is the angle of the steering direction, relative to the perpendicular vector from the antenna.

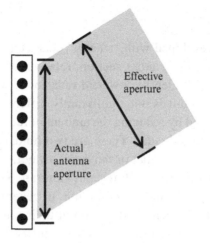

Figure 16.4

16.3 Radar Range Equation

Detection of objects using radar involve sophisticated signal processing. However, all this is first dependent on the amount of energy received from the target echo:

$$\text{Receiver power } P_{\text{receive}} = P_t G_t A_r \sigma F^4 (t_{\text{pulse}}/T)/((4\pi)^2 R^4)$$

where

P_t = transmitted power
G_t = antenna transmit gain
A_r = receive antenna aperture area
σ = radar cross-section (a function of target geometric cross-section, reflectivity of surface, and directivity of reflections)
F = pattern propagation factor (unity in vacuum, accounts for multipath, shadowing, and other factors)
t_{pulse} = duration of receive pulse
T = duration of transmit interval, or the inverse of the PRI
R = range between radar and target

Notice that the received power drops with the fourth power of the range, so radar systems must cope with very large dynamic ranges in the receive signal processing. The radar energy seen by the target drops proportional to the range squared. The reflected energy seen by the radar receive antenna further drops by a factor of the range squared. The capability to detect very small signals is crucial to operate at longer ranges.

16.4 Stealth Aircraft

Military planes have been developed with "stealth" characteristics. This means that such a plane has a very small σ, or radar cross-section, relative to other aircraft of similar size. It can still be detected by a sufficiently powerful radar or at sufficiently close ranges. Because the size of stealth aircraft is not significantly different from other planes, the stealth characteristic is achieved by reducing the amount of radar signal that is reflected back from the aircraft to the transmitted radar. There are two fundamental ways to reduce these reflections: either absorb the radar signal or deflect it away from the radar transmitter. Special radar-absorbent materials are used in stealth aircraft. The shape and contours of the aircraft also greatly influence the radar cross-section. In general, a concave surface tends to reflect radar waves back in the general direction of their source, which is not good for a stealth plane. Examples of concave surfaces are engine inlets, right angles where wings join the fuselage, open bomb bays, and even the cockpit if the windscreens are transparent to radar waves. Convex surfaces, on the other hand, tend to scatter radar waves in widely separated directions, reducing the amount of signal reflected back to the source. For example, the B2 stealth bomber is shaped like a flying wing, which is basically a convex shape when viewed from just about any direction. Smaller features, such as the engine air inlets, have a geometry designed to reflect impinging radar signals in a direction other than that of the illuminating radar.

16.5 Pulsed Radar Operation

Most radar systems are pulsed, meaning that the radar transmits a pulse and then listens for receive signals, or echoes. This type of system avoids the problem of a sensitive receiver trying to operate simultaneously with a high-power transmitter. The pulse width or duration is an important factor. The shorter the pulse width, the easier it is to determine range because the receive signal is of short duration also. Radars operate by "binning" the receive signals. The receive signal returns are sorted into a set of bins by time of arrival relative to the transmit pulse. This is proportional to the roundtrip distance to the object(s) reflecting the radar waves. By analyzing the receive signal strength in the bins, the radar can sort the objects by radar cross-section size and across different ranges. This analysis is performed for all desired azimuths and elevations.

Having many range bins allows more precise range determinations. A short duration pulse is likely to be detected and mapped into only one or two range bins, rather than being spread over many bins. However, a longer pulse duration or width allows for a greater amount of signal energy to be transmitted, and a longer time for the receiver to integrate the energy. This means longer detection range. To optimize for both fine-range resolution and long-range detection, radars use a technique called pulse compression.

16.6 Pulse Compression

The goal of pulse compression is to transmit a long duration pulse of high energy, but to detect a short duration pulse to localize the receive filter output response to one or at most two radar range bins. Early radars accomplished this by transmitting a signal with linear frequency modulation. The pulse would start at a low-frequency sinusoid and increase the frequency over the duration of the radar pulse. This is referred to as a "chirp." A special analog filter is used at the receive end, with a nonlinear phase response. This filter has a time lag that decreases with frequency. When this rate of time lag decrease is matched to the rate of increase in the chirp, the result is a very short, high-amplitude output from the filter. The response of the pulse detection has been "compressed."

All digital radars can also perform pulse compression, but using a different method. Recall the matched filter in Chapter 9 on complex modulation and demodulation. The matched filter performs the same effect as the analog pulse compression technique just described. If the transmitted radar pulse uses a pseudorandom sequence of phase modulations and is detected using a matched filter, then the resulting output is of high amplitude only when the receive signal sequence matches up in phase (or delay) to the transmitted pulse sequence. This approach can be used to precisely identify delay or time of arrival of the receive pulse. The sequence used for radar transmit pulses must have strong autocorrelation properties (sequence of length N correlates to value N with zero offset, and to 0 for any nonzero sequence offset). In radar systems, sequences known as Barker sequences are sometimes used.

16.7 Pulse Repetition Frequency

A high pulse repetition frequency (PRF) has several advantages. First, the higher the PRF, the greater the average power the radar is transmitting (assuming the peak power of each pulse is limited by the transmit circuitry), and the better the chance of detection of targets. A high or fast PRF also allows for more rapid detection and tracking of objects because range measurements at a given azimuth and elevation can be performed during each PRF interval. A high PRF also allows easier discrimination of the Doppler frequency, a topic discussed in the next chapter. But a low PRF also has an important advantage, which is to allow unambiguous determination of range over longer distances. This is our next topic.

Range to target is measured by roundtrip delay in the received echo. It is the speed of light multiplied by the time delay and divided by two to account for the roundtrip:

$$R_{measured} = v_{light} t_{delay}/2$$

The maximum range that can be unambiguously detected is limited by the PRF. This is more easily seen by example. If the PRF is 10 kHz, then we have 100 µs between pulses.

Therefore, all return echoes should ideally be received before the next transmit pulse. This range is simply found by multiplying the echo delay time by the speed of light and dividing by two to account for the roundtrip:

$$R_{maximum} = (3 \times 10^8 \text{ m/s})(100 \times 10^{-6} \text{ s})/2 = 15 \text{ km}$$

Let us suppose the radar system sorts the returns into 100 range bins, based on the time delay of reception. The range resolution of this radar system is then 0.15 km, or 150 m. However, there may be returns from distances beyond 15 km. Suppose that a target aircraft 1 is 5 km distant, and a target aircraft 2 is 21 km distant. Target aircraft 1 has a delay of

$$t_{delay} = 2R_{measured}/v_{light} = 2(5 \times 10^3)/(3 \times 10^8) = 33 \text{ μs}$$

Target aircraft 2 has a delay of

$$t_{delay} = 2R_{measured}/v_{light} = 2(21 \times 10^3)/(3 \times 10^8) = 140 \text{ μs}$$

The first target return is mapped into range bin 33 out of 100, and the second target to range bin 40. This is called a range ambiguity. The target or targets that are within the 15 km are said to be in the unambiguous range. This is analogous to the sampling rate. The range ambiguity is analogous to aliasing during the sampling process (see Figure 16.5).

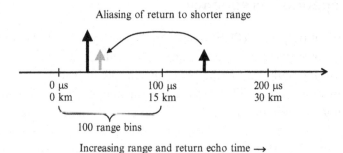

Figure 16.5

One solution to this problem is to transmit different pulses at each PRF interval. However, this approach has the downside of complicating the receiver because it must now use multiple matched filters at each range bin and at each azimuth and elevation. Doing so effectively doubles the rate of digital signal processing required for each separate transmit pulse and matched filter pair used.

Another approach can be used instead. If we periodically change the PRF slightly, we find that the returns in the unambiguous range do not move. However, those beyond that range shift in their apparent ranges. This approach can be illustrated using an example.

Suppose we switch the PRF to 11 kHz, from 10 kHz, or 90.9 μs. The maximum unambiguous range is

$$R_{maximum} = (3 \times 10^8 \text{ m/s})(90.9 \times 10^{-6} \text{ s})/2 = 13.6 \text{ km}$$

The target aircraft at 5 km distance still has a 33 μs delay. If we use 100 bins like before, the target return appears in bin number $100 \times 33 / 90.9 = 36$th bin. The target aircraft 2 at 21 km has a target return delay of 140 μs. This appears as a return at $140 - 90.9 = 49.1$ μs, and in $100 \times 49.1 / 90.9 = 54$th bin.

So by switching PRFs, we are able to determine that at least one of our targets is beyond the ambiguous range:

  PRF = 10 kHz: Target aircraft returns in bin 33 and in bin 40.
  PRF = 11 kHz: Target aircraft returns in bin 36 and in bin 54.

We assume Scenario A:

  Target1 moved from bin 33→36 when we changed PRF.
  Target2 moved from bin 40→54 when we changed PRF.

Instead, what if we assume Scenario B:

  Target1 moved from bin 33→54.
  Target2 moved from bin 40→36.

From this information, we cannot be sure that the first target was at 5 km and the second at 21 km. We do not work this out here, but if Target1 is at 34.8 km (or 232 μs) and Target2 is at 141 km (or 940 μs), the result is Scenario B.

The way to tell which scenario is, in fact, occurring is to use a third PRF, perhaps at 9 kHz, or 111 μs:

$$R_{maximum} = (3 \times 10^8 \text{ m/s})(111 \times 10^{-6} \text{ s})/2 = 16.7 \text{ km}$$

The target aircraft at 5 km appears in bin number $100 \times 33/111 = 30$th bin. The target aircraft 2 at 21 km appears in bin number $100 \times (140 - 111)/111 = 26$th bin.

This additional information allows us to know that Scenario A is the true one.

In reality, there may be many target returns, and they may also be obscured by noise or clutter in the return. The higher the PRF, the more ambiguity is present in the range returns. For these reasons, radar detection is at best a statistical process, with calculated probability of detections (at a given range for a given radar cross-section). There is also the probability of false detection that must be considered when setting detection thresholds.

16.8 Detection Processing

Most radars have thousands of range bins. They may scan wide sweeps of azimuth and elevation. Or in tracking mode, they may be focused in narrow regions containing targets that have been detected. In either case, the rate of digital signal processing can be very high.

A matched filter can be used to detect incoming radar pulses. The radar focuses at a particular azimuth and elevation for one or many PRFs. For each PRF, the incoming data is filtered using an FIR filter with an impulse response that is the complex conjugate of the radar transmit pulse. This produces a large peak in the filter output at the point where the incoming data stream contains radar pulse, which corresponds to a particular range bin.

The computation load in modern radars can be very high.

Filtering or convolving the receive signals by using matched filters is a computationally intense process. This could be done using FIR filters. However, in order to reduce the amount of computations, another method is often used.

An alternative is to perform an FFT transform of the receive signal sequence from each PRF. The spectral representation of the receive signal can then by multiplied by the frequency response of the radar pulse. After this multiplication, the result can either be transformed back into the time domain using an IFFT or all subsequent processing performed in the frequency domain. This process performs the equivalent function as FIR filtering. This process sounds counterintuitive, but due to the efficiency of the FFT algorithm, it is often less computationally intensive than a large FIR filter. In any case, once the receive pulse has been processed, the amplitude of the matched filter operation is compared to a threshold to determine if this is a valid radar pulse return. Results over multiple PRFs can be used to discard or confirm valid target radar returns, maximize the probability of detection, and minimize the probability of false detections.

In the next chapter, we discuss Doppler processing. This also requires use of the FFT algorithm. In radar systems, the FFT is the most common digital signal processing algorithm, and efficient implementation of FFTs is critical for any digital radar system.

CHAPTER 17

Pulse Doppler Radar

We mentioned Doppler frequency shift in our discussions of wireless systems. In general, these frequency shifts degrade wireless receiver performance and must be compensated for. In radar, however, Doppler shifts are a key part of the detection and tracking of objects. For this reason, nearly all radar systems incorporate Doppler processing.

By measuring the Doppler rate, the radar is able to measure the relative velocity of all objects returning echoes to the radar system—whether planes, vehicles, or ground features. For targeting radars, estimating the target's velocity is equally as important as determining its location. And for all radars, Doppler filtering can be used to discriminate between objects moving at different relative velocities. This capability can be especially important when there is a high level of clutter obscuring the target return. An example of this might be an airborne radar trying to track a moving vehicle on the ground. Since the ground returns are at the same range as the vehicle, the difference in velocity is the means of discrimination using Doppler measurements.

17.1 Doppler Effect

Because sensing Doppler frequency shifts is so important, it is worth reviewing their causes. A common example we have all experienced is standing beside a train track or highway. As a train or truck approaches, we hear a certain frequency sound. As a high-speed train or truck passes, the sound immediately drops several octaves. This drop is caused by a frequency shift called the Doppler effect. Although we cannot sense this, the light waves are affected in the same way as the sound waves. In fact, the realization that our universe is expanding was determined by making very fine Doppler measurements of the light from stars in the night sky.

The relationship between wavelength and frequency is

$$\lambda = v/f$$

where

    f = wave frequency (Hz or cycles/s)
    λ = wavelength (m)
    v = speed of light (approximately 3×10^8 m/s)

Digital Signal Processing 101. DOI: 10.1016/B978-1-85617-921-8.00021-3

The speed of light is constant—Einstein proved this. Technically, this is true only in a vacuum, but the effect of the medium, such as our atmosphere, can be ignored in radar discussions. In a radar system the frequency is modified by the process of being reflected by a moving object. Consider the transmission of a sinusoidal wave. The distance from the crest of each wave to the next is the wavelength, which is inversely proportional to the frequency. Each successive wave is reflected from the target object of interest. When this object is moving toward the radar system, the next wave crest reflected has a shorter roundtrip distance to travel, from the radar to the target and back to the radar. The reason is that the target has moved closer in the interval of time between the previous and current wave crest. As long as this motion continues, the distance between the arriving wave crests is shorter than the distance between the transmitted wave crests. Since frequency is inversely proportional to wavelength, the frequency of the sinusoidal wave appears to have increased. If the target object is moving away from the radar system, then the opposite happens. Each successive wave crest has a longer roundtrip distance to travel, so the time between arrival of receive wave crests is lengthened, resulting in a longer (larger) wavelength and a lower frequency. This effect becomes more pronounced when the frequency of the transmitted sinusoid is high (short wavelength). Then the effect of the receive wavelength being shortened or lengthened due to the Doppler effect is more noticeable. Therefore, Doppler frequency shifts are more easily detected when using higher frequency waves because the percentage change in the frequency is larger.

This effect applies only to the motion relative to the radar and the target object. If the object is moving at right angles to the radar, there is no Doppler frequency shift. An example of this would be an airborne radar directed at the ground immediately below an aircraft. Assuming the terrain is level and the aircraft is at a constant altitude, the Doppler shift is zero, even though the plane is moving relative to the ground. There is no change in the distance between the plane and ground.

If the radar is ground based, then all Doppler frequency shifts are due to the target object motion. If the radar is vehicle or airborne based, then the Doppler frequency shifts are due to the relative motion between the radar and target object. For example, if you are driving on the highway at 70 mph, and an approaching police car is traveling at 50 mph, the radar shows a Doppler shift corresponding to 120 mph. The police radar needs to subtract the speed of the police car to display your speed.

This capability can be of great advantage in a radar system. By binning the receive echoes both over range and Doppler frequency offset, the radar can determine target speed as well as range. Also, this allows easy discrimination between moving objects, such as an aircraft, and the background clutter, which is generally stationary.

Imagine we have a radar operating in the X band at 10 GHz ($\lambda = 0.03$ m or 3 cm). The airborne radar, traveling at 500 mph, is tracking a target ahead moving at 800 mph in the same direction. In this case, the speed differential is −300 mph, or −134 m/s.

Another target is traveling head-on toward the airborne radar at 400 mph. This gives a speed differential of 900 mph, or 402 m/s The Doppler frequency shift can be calculated as follows:

$$f_{Doppler} = 2v_{relative}/\lambda$$

$$\text{First target Doppler shift} = 2(-134 \text{ m/s})/(0.03 \text{ m}) = -8.93 \text{ kHz}$$

$$\text{Second target Doppler shift} = 2(402 \text{ m/s})/(0.03 \text{ m}) = 26.8 \text{ kHz}$$

The receive signal is offset from 10 GHz by the Doppler frequency. Notice that the Doppler shift is negative when the object is moving away (opening range) from the radar and is positive when the object is moving toward the radar (closing range).

17.2 Pulsed Frequency Spectrum

Measuring Doppler shift is complicated by the fact that the radar is transmitting pulses. This has an effect on the spectrum of the radar transmit signal. To understand this, we need to start with the frequency, or spectral representation, of a pulse. In Chapter 11 on digital upconversion, we discussed the frequency response of DACs. The time response of a DAC is also pulse, and we saw that the frequency response in the sin(x)/x or sinc function. If the pulse has the sharp edges removed, we can reduce the side lobes, although doing so broadens the main lobe.

The frequency response of an infinite train of pulses is similar, except that it is composed of discrete spectral lines in the envelope of the sinc function (see Figures 17.1–17.3). Also, the spectrum repeats at intervals of the pulse repetition frequency (PRF). We forgo the mathematical derivation here, but it is available in any engineering text on radar. This is not unlike a sampled signal, in which the frequency representation repeats at the sampling frequency interval.

The important point is that the spectral line spacing imposes restrictions on Doppler frequency shifts. For the radar to unambiguously identify the Doppler frequency shift, it must be less than the PRF frequency. Doppler frequency shifts greater than this alias to a lower

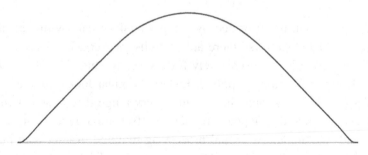

Figure 17.1:
Spectrum of single pulse.

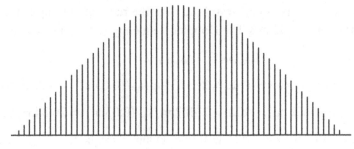

Figure 17.2:
Spectrum of a pulse train repeating slowly.

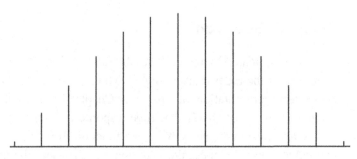

Figure 17.3:
Spectrum of a pulse train repeating rapidly.

Doppler frequency and are ambiguous, just as radar returns beyond the range of the PRF interval time are ambiguous because they alias into lower range bins.

The Doppler frequency range placement is somewhat arbitrarily determined by the digital downconversion of the received radar high-frequency carrier to baseband. Assuming downconversion of the carrier to 0 Hz, then the Doppler frequency effect causes the target return signal to have a positive or negative offset, as computed here:

$$f_{Doppler} = 2v_{relative}/\lambda.$$

Doppler frequency detection is performed by using a bank of narrow digital filters, with overlapping frequency bandwidth (so there are no nulls, or frequencies that could go undetected). This detection is done separately for each range bin. Therefore, at each allowable range, Doppler filtering is applied. Just as the radar looks for peaks from the matched filter detector at every range bin, within every range it tests across the Doppler frequency band to determine the Doppler frequency offset in the receive pulse. This process dramatically expands the amount of signal processing required. Rather than build many individual narrow-band frequency filters, the radar uses the FFT to perform spectral filtering across the spectral bandwidth of the PRF signal.

17.3 Doppler Ambiguities

Doppler ambiguities can occur if the Doppler range is larger than the PRF. The maximum Doppler requirement of a given radar can be estimated. Let's use a military airborne radar example. The fastest closing rates are with approaching targets because speeds of both the radar-bearing aircraft and the target aircraft are summed. This should assume the maximum speed of both aircraft. The highest opening rates might be when a target is flying away from the radar-bearing aircraft. Here, we should assume the radar-bearing aircraft is traveling at minimum speed, and the target aircraft is flying at maximum speed. We should also assume the target aircraft is flying a large angle θ from the radar-bearing aircraft flight path, which further reduces the radar-bearing aircraft speed in the direction of the target.

Maximum positive Doppler frequency (fastest closing rate) at 10 GHz or 3 cm
Radar-bearing aircraft maximum speed: 1200 mph = 536 m/s
Target aircraft maximum speed: 1200 mph = 536 m/s
Maximum positive Doppler = 2 (1072 m/s) / (0.03 m) = 71.5 kHz

Maximum negative Doppler frequency (fastest opening rate) at 10 GHz or 3 cm
Radar-bearing aircraft minimum speed: 300 mph = 134 m/s
Effective radar-bearing aircraft minimum speed with θ = 60-degree angle from target track is sin(60) = 0.5: 150 mph = 67 m/s
Target aircraft maximum speed: 1200 mph = 536 m/s
Maximum negative Doppler = 2 (67 − 536 m/s) / (0.03 m) = −31.3 kHz

This gives a total Doppler range of 71.5 + 31.3 = 102.8 kHz. Unless the PRF exceeds 102.8 kHz, there is aliasing of the detected Doppler rates and the associated ambiguities.

Below, in Figure 17.4, the aliasing resulting in Doppler ambiguity is shown for a higher PRF of 80 kHz. If the PRF was 10 kHz, there would be many more Doppler ambiguities in the spectrum.

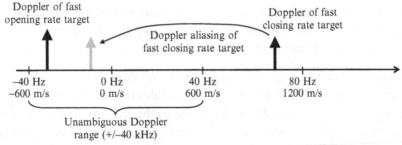

Example : PRF = 80 kHz, with 10 GHz radar

Figure 17.4:
Doppler frequency diagram

If we assume a PRF of 10 kHz from the preceding chapter's example, we clearly have Doppler ambiguities. Doppler ambiguities can be resolved using a number of methods:

- Range Differentiation: Using range measurements over a period of time, we can measure the difference in range over the time interval. Using this, the radar can estimate the change in range, which is the relative velocity between the radar and the target. This method is less precise than Doppler-based measurements but can provide an estimate to use in resolving the Doppler ambiguity.
- Multiple or offset PRFs: This method is very similar to resolving range ambiguities. Multiple PRFs with slightly different values can be used, and the ambiguities resolved by analysis of how the aliased Doppler frequency measurements move within the unambiguous range.
- Variable PRF: The PRF need not be constant, particularly in a digitally programmable system. The PRF can be varied. The PRFs are generally grouped into low, medium, or high ranges. A low PRF is generally from 2 to 8 kHz. A medium PRF is generally from 8 to 30 kHz. And a high PRF is generally from about 30 to 250 kHz. Each PRF zone has its advantage and disadvantages.

We have already discussed range and Doppler ambiguities. The PRF directly affects the size of the unambiguous zone. But ambiguities are not the only issue. Just as a range or Doppler measurement return is outside the unambiguous zone and is aliased into the primary zone, so are all other returns and radar clutter. This can raise the noise floor of the radar to such a degree that lower amplitude returns become obscured.

17.4 Radar Clutter

There are two categories of radar clutter: main lobe clutter and side lobe clutter. Main lobe clutter occurs when there are undesirable returns in the main lobe or within the radar beamwidth. This usually occurs when the main lobe intersects the ground. This situation can occur because the radar is aimed downward (negative elevation); there is higher ground such as mountains in the radar path; or even if the radar beam is aimed level, and as the beam spreads, it hits the ground. Because the area of ground in the radar beam is often large, the return can be much larger than targets.

Side lobe clutter is unwanted returns coming from a direction outside the main lobe. Since the radar is not pointed in this direction, it is never a desired radar return signal. Side lobe clutter is usually attenuated by 50 dB or more, due to the antenna directional selectivity or directional radiation pattern. A very common source of side lobe clutter is ground return. When a radar is pointed toward the horizon, there is a very large area of ground covered by the side lobes in the negative elevation region. The large reflective area covered by the side lobe can cause significant side lobe returns despite the antenna attenuation.

Different types of terrain have a different "reflectivity," which is a measure of how much radar energy is reflected back. This measure also depends on the angle of the radar energy relative to the ground surface. A related parameter, known as the backscattering coefficient, has the angle incorporated into the coefficient and is therefore normalized over all angles. Some surfaces, like smooth water, reflect most of the radar energy away from the radar transmitter, particularly at shallow angles. A desert would reflect more of the energy back to the radar, and wooded terrain would reflect even more. Manmade surfaces, such as in urban areas, would reflect the most energy back to the radar system.

This reflectivity is one reason why Doppler processing is so important. Most targets are moving, and Doppler processing is an effective method to distinguish them from the background clutter of the ground. Remember, the Doppler frequency of the ground is usually nonzero if the radar is in motion. In fact, side lobe Doppler clutter varies by the elevation and azimuth angle because the relative velocity is proportional to cosine θ, where θ is equal to the angle between the aircraft flight line and a given location on the ground.

Different points on the ground, depending on how far ahead or off to the side of the radar-bearing aircraft that particular patch of ground is located, will have different relative velocities and therefore Doppler frequencies.

So Doppler side lobe clutter is present over a wide range of Doppler frequencies.

Main lobe clutter is more likely to be concentrated at a specific frequency, since the main lobe is far more concentrated (typically 3-6 degrees of beam width), so the patch of ground illuminated is likely to be far smaller and all the returns at or near the same relative velocity.

A simple example can help illustrate how the radar can combine range and Doppler returns to obtain a more complete picture of the target environment.

The diagram in Figure 17.5 illustrates unambiguous range and Doppler returns. This assumes the PRF is low enough to receive all the returns in a single PRF interval, and the PRF is high enough to include all Doppler return frequencies.

The ground return comes though the antenna side lobe, known as side lobe clutter. Ground return is often high due to the amount of reflective area at close range, which results in a strong return despite the side lobe attenuation of the antenna. The ground return is at short range, essentially the altitude of the aircraft. In the main lobe, the range return of the mountains and closing target are close together, due to similar ranges. It is easy to see how if we are just using the range return, a target return can be lost in high-terrain returns, known as main lobe clutter.

The Doppler return gives a different picture. In this case, the ground return is more spread out, around 0 Hz. The ground slightly ahead of the radar-bearing plane is at a slightly positive

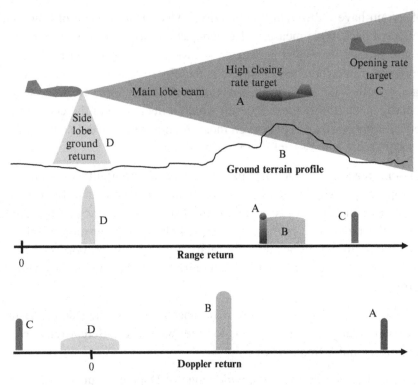

Figure 17.5

relative velocity, and the ground behind the plane is at a slightly negative relative velocity. As the horizontal distance from the radar-bearing plane increases, the ground return weakens due to increased range.

Notice the Doppler return from the mountain terrain is now very distinct from the nearby closing aircraft target. The mountain terrain is moving at a relative velocity equal to the radar-bearing plane's velocity. The closing aircraft's relative velocity is the sum of both aircrafts' velocities, which are very high, producing a Doppler return at a high velocity. The other target aircraft, which is slowly opening the range with the radar-bearing aircraft, is represented as a negative Doppler frequency return.

17.5 PRF Trade-offs

Different PRF frequencies have different advantages and disadvantages. The following discussion summarizes the trade-offs.

Low PRF operation is generally used for maximum range detection. It usually requires a high-power transmit power to receive returns of sufficient power for detection at a long range (remember, receive echo power levels are proportional to the range to the fourth power). To get the highest power, the radar system sends long transmit pulses, and correspondingly long-matched filter processing (or pulse compression) is used. This mode is useful for precise range determination. Strong side lobe returns can often be determined by their relatively close ranges (ground area near the radar system) and filtered out. Disadvantages are that Doppler processing is relatively ineffective due to so many overlapping Doppler frequency ranges. This limits the ability to detect moving objects in the presence of heavy background clutter, such as moving objects on the ground. These overlapping ranges can also be a problem for detecting low-flying aircraft because of ground terrain clutter at similar ranges in the main lobe of the radar.

High PRF operation spreads out the frequency spectrum of the receive pulse, allowing a full Doppler spectrum without aliasing or ambiguous Doppler measurements. The clutter that is present in the spectrum is not folding or aliased from higher frequencies, which lowers the noise floor of the receive spectrum. A high PRF can be used to determine Doppler frequency and therefore relative velocity for all targets. It can also be used when a moving object of interest is obscured by a stationary mass, such as the ground or a mountain, in the radar return. The unambiguous Doppler measurements make a moving target stand out from a stationary background. This is called main lobe clutter rejection, or filtering. Another benefit is that since more pulses are transmitted in a given interval of time, higher average transmit power levels can be achieved. This can help improve the detection range of a radar system in high PRF mode.

Pulse delay-based ranging performance becomes very compromised in high PRF operation. One solution is to use the high PRF mode to identify moving targets, especially fast-moving targets, and then switch to a low PRF operation to determine range. Another alternative is to use a technique called FM ranging. In this mode, the transmit duty cycle becomes 100%—the radar transmits continuously. But it transmits a continuously increasing frequency signal and then at the maximum frequency abruptly begins to transmit at a continuously decreasing frequency until it reaches the minimum frequency whereby it resets to begin another cycle of increasing frequency. The frequency over time looks like a "saw-tooth wave." The receive can operate during transmit operation because the receiver is detecting time-delayed versions of the transmit signal, which are at a different frequency than current transmit operation. Therefore, the receiver is not desensitized by the transmitter's high power at the received signal frequency.

The receiver can continue to operate during transmit operation, as the receiver is detecting time delayed versions of the transmit signal, which are at a different frequency than simultaneous transmit frequency. Since the receive and transmit frequencies are different at any given moment, the receiver is not desensitized by the transmitter's high power at the received signal frequency. Measurement of the receive frequency, and knowing the transmitter frequency ramp timing, can be used to detect roundtrip delay time, and therefore range.

This method is not as accurate as pulse delay ranging using a matched filter but can provide ranging information nonetheless. Of course, the receive frequency is affected by the Doppler frequency. On a rapidly closing target, the receive frequencies are all offset by a positive $f_{Doppler}$, which can be measured by the receiver once the peak receive frequency is detected. The Doppler addition can be found because the receiver knows the peak frequency of the transmitter (see Figure 17.6).

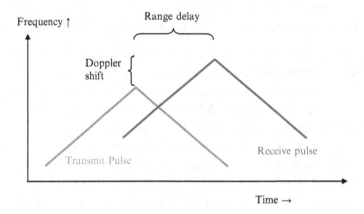

Figure 17.6

Medium PRF operation is a compromise. Both range and Doppler measurements are ambiguous, but neither is aliased or folded as severely as the more extreme low or high PRF modes. This operation can provide a good overall capability for detecting both range and moving targets. However, the folding of the ambiguous regions can also bring a lot of clutter into both range and Doppler measurements. Small shifts in PRFs can be used to resolve ambiguities, as has been discussed, but if there is too much clutter, the signals may be undetectable or obscured in both range and Doppler.

17.6 Target Tracking

So far, we have discussed how a radar system performs measurements of range and velocity of potential targets. After they are measured, the targets can be identified using some of the methods already described. A target may have a specific return amplitude, azimuth, elevation, range, and relative velocity. Since measurements are repeated continuously, this allow for tracking of targets.

To track an identified target, the radar system uses repeated measurements over time. These measurements can be filtered to reduce measurement error and the results of the filtering fed back to control the measurement process. The radar system can respond by aiming the main lobe, by changing PRF, and by using measurements to anticipate future behavior of the target. For example, by estimating the target velocity and knowing the lag or latency in measurements, the radar can estimate the next position of the target and have the main lobe lead the target motion. This can also be done for the range binning and Doppler filtering. Also, if the radar itself is on a moving platform, such as an aircraft, the motion of the radar-bearing platform needs to be taken into account. This is referred to as platform stabilization.

Filtering of target measurements can be much more complex than the basic digital filtering discussed in previous chapters. The filtering may be recursive, where previous filter outputs are fed back, and is adaptive, with gains and frequency cutoffs varied in response to the measurement accuracy, degree of clutter, angle of antenna main beam, and other factors. There may be a number of independent or dependent filtering loops in operation. One loop may be tracking the range of a target, by monitoring the range bins and by detecting the comparative changes in adjacent range bin results, known as range gating. By doing this, the radar system can coarsely estimate the rate of change of the range, which can lead to a decision on when to switch to a high PRF to confirm with Doppler measurements. The antenna main lobe may be electronically steered by making measurements at elevation and azimuths slightly above/below or side to side to estimate the azimuth and elevation of the highest return, to keep the main lobe centered on a target. This process, known as angle tracking, also must account for motion of the radar platform. Note that this tracking activity may be a portion of the time, whereas another portion of the time can be used for scanning or tracking of other targets.

In addition to tracking, there are often software-implemented algorithms to correlate the various measurements to particular targets. Target ranges, velocities, azimuths, and elevations may cross over each other. These changes need to be interpolated into a trajectory that can be matched to a specific target. Using radar digital signal processing followed by software-enabled target identification, tracking, monitoring, and classification, the radar system can automate all these functions. Higher-level tracking by software can also allow for improvements in probability of detection and minimization of false alarms because behavior of potential targets can be correlated and analyzed over longer time intervals than the radar measurement functions typically performed. This can allow the operator or pilot to more quickly understand the situation and spend more time on deciding how to respond.

CHAPTER 18

Synthetic Array Radar

Synthetic Array Radar, or SAR, is normally used to map ground features and terrain. It is also known in the literature as Synthetic Aperture Radar. Both names make sense, though we use Synthetic Array Radar here. This type of radar is used for a wide variety of military and commercial applications. It can be made to map almost arbitrarily fine resolution ground features or used to more coarsely map larger areas with comparative effort.

This process produces maps, which are often color coded. The color does not represent the actual color of the landscape but is used to indicate the strength of return signals for each resolvable location on the ground. Alternately, the images can be grayscale, with light regions indicating strong return and dark regions indicating little or no return. Because different terrain features reflect radar in differing amounts, features such as buildings, planes, rivers, roads, railroads, and so on can be seen in the SAR images.

18.1 SAR Resolution

The key parameter in ground mapping is the resolution. SAR systems can be designed with capabilities to differentiate using dimensions from few centimeters to hundreds of meters, depending on if the purpose is to map out a military installation, an urban area, or the contours of a mountain range. The range is basically limited by the transmit power of the radar, which can operate at resolutions much greater than visual detection at long ranges, and is unaffected by darkness, haze, and other factors impacting visual detection.

As with video, the quality of images depends on the pixel density (pixel stands for picture element). The equivalent of pixel density in radar is a voxel, or volume element. The voxel is defined by the azimuth, elevation, and range. The minimum voxel size is dependent on the radar resolution capabilities. The voxel spacing is basically the distance that two points on the ground can be distinguished from each other. Radar resolution capabilities, in turn, are dependent on range resolution and main lobe beamwidth capabilities.

The voxel spacing or density should, in general, be at least 10 times the dimensions of the objects being mapped to achieve useful images. A 1 m resolution is feasible for detecting buildings that are at least 10 m long and wide.

Digital Signal Processing 101. DOI: 10.1016/B978-1-85617-921-8.00022-5

Because precision range detection is a fundamental requirement, high PRF operation is unsuitable for SAR due to the range ambiguities. Low PRFs are used instead, to eliminate range ambiguities over the distances from the aircraft to the ground being mapped. Maximum Doppler rates tend to be low because the only motion is due to the radar-bearing aircraft motion. Due to the nature of SAR, the relative motion is normally substantially less than the aircraft flight speed. Use of a low PRF, while restricting the usable Doppler range, enhances the precision of Doppler frequency detection within that restricted range. This is an advantage in high-resolution SAR mapping.

18.2 Pulse Compression

Range resolution is dependent on the precision of the receive pulse detection arrival delay. This resolution can be achieved by a very short transmit pulse width, which has the disadvantage of a low transmit power level due to the short duration. Or very high levels of pulse compression can be used, which allow relatively long transmit pulses and therefore long integration times at the receiver, with the receiver operating on higher-power returns. This raises the SNR and allows for longer-range mapping. A high level of pulse compression can be achieved by using long-matched filters (a correlation to the complex conjugate of transmit sequence) and transmit sequences with strong autocorrelation properties. The only consequence is a higher level of computations associated with the long-matched filter. The speed of light, and therefore radar waves, is about 1 m per 3 ns (3×10^{-9}). Since the path is roundtrip, the range appears to become half this. So about 2 m, or $\sim$ 6 feet range resolution, requires a 12-ns timing detection precision. To achieve this level of correlation would require a transmit sequence with phase changes of at least 80 MHz rate, resulting in, at a minimum, the same amount of transmit frequency bandwidth within a typical 10 GHz radar band.

The elevation of the antenna main lobe does not need to be narrowly focused. In an SAR radar system, the antenna is directed to the ground at an angle off to the side, as shown in Figure 18.1. As the elevation angle decreases, the radar beam is directed at a steeper angle to a ground location closer to the flight path of the aircraft, with a shorter range. The different portions of the beam elevation therefore map to different ranges, and the return sequence can be directed into different range bins. The precision of the range detection capability translates into the degree of elevation resolution, often utilizing pulse compression.

18.3 Azimuth Resolution

The other requirement for precise ground mapping is for a very narrow angular resolution of the main beam in the azimuth. As discussed in a previous chapter, the narrowness of the radar

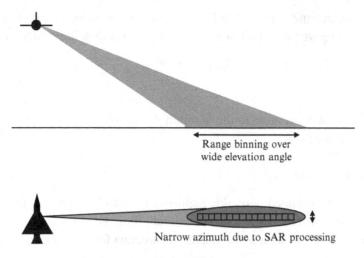

Figure 18.1

beam depends on the ratio of the antenna size to the wavelength. To achieve a "pencil-like" radar beam requires either a very large antenna or very high-frequency (and short wavelength) radar. Both are impractical for airborne radar. The antenna size is necessarily limited by the aircraft size. Extremely high-frequency radars tend to be useful only at very short range, due to both atmospheric absorption and scattering. There is also the practicality of building high-power and very high-frequency transmit circuits.

The solution to this problem is to create an artificially large antenna, or synthetic array antenna. The forward motion of the aircraft is used to transmit and receive from many different points along the flight path of the aircraft. When the radar main beam is focused at the same area of ground during the aircraft motion, the returns from different angles created by the aircraft motion can be synthesized into a very narrow equivalent azimuth main lobe using signal processing techniques. The end result is as if an antenna of great length (up to a kilometer) was used. Because this technique is done using radar returns over several hundred milliseconds, it works for stationary targets and so is ideal for ground mapping.

SAR radar typically directs a radar beam at 90° to the plane's flight path. The width of this radar beam does not have to be exceptionally narrow; in fact, the wider beam covers more ground and allows more processing gain in the SAR algorithm. When a large-angle main lobe is used, the maximum length of the synthetic antenna is increased. Therefore, small antennas can work well with the SAR technique, as long as the antenna gain is sufficient to meet the SNR requirements for the range involved. The antenna illuminates a large swath on the ground, typically an oval shape due to the aspect ratio of the beam being aimed outward from the aircraft flight path at a downward angle.

To start with, let us assume that we can build an antenna as large as necessary to meet our azimuth resolution requirement. The rule of thumb governing antenna size is

$$d_{azimuth} \approx \lambda R/L$$

where

$d_{azimuth}$ = Resolvable distance in the azimuth direction
λ = Wavelength of radar
R = Range
L = Length of the antenna

(Note, for reasons not explained here, this expression is valid for conventional antennas. An SAR antenna actually has half the resolvable azimuth limit as a real antenna. In other words, an SAR antenna needs to be twice as large as a real antenna for the same resolvable distance.)

If we need a 1-m aperture at a 10-km range, with a 3-cm (X band) radar, this requires an antenna length of 300 m.

Imagine we had such an antenna, mounted along the fuselage of an impractically long 300 m long plane. Each pulse could be focused with an azimuth width of 1 m at the 10 km range, with a wide elevation, allowing the radar to scan a narrow (1 m) strip of land each PRF as the plane travels forward.

This 300 m long antenna could be composed of many radiating elements along this length (e.g., 301 separate elements, spaced every meter). The antenna steering is accomplished by setting the phase of each element to ensure that the radar wave transmitted from each element is at a common phase when arriving at the 1 m strip at a 10-km distance. The phases have to be adjusted, or focused, because the distance to the 1 m strip of land is slightly different for each element, due to the offset relative from the center element.

To aim the antenna beam at a very narrow region, we must carefully control the phase relationship of the different antenna elements. In our example, the wavelength is 3 cm. If a radar roundtrip path is 1.5 cm longer or shorter than the middle element path, it is 180° out of phase, and adds destructively or cancels. Therefore, the roundtrip path length must be controlled within a few millimeters for each element. We must compensate for the phase error that occurs due to the plane's straight flight path as compared to an arc. This is shown below. The phase correction relative to the middle element of the antenna works is approximately

$$\Theta_n = (2 \cdot \pi \cdot d_n^2)/(\lambda \cdot R)$$

where

Θ_n = Phase error of nth antenna element (in radians)
d_n = Distance between middle element and nth antenna element

(In SAR literature, the term Θ_n is sometimes called the "point target phase history.")

The return echo would be reflected from the ground at all locations and travel back to each element. Due to the reciprocal path, it would arrive at the same phase offset that was transmitted, and if the same phase compensation is performed on the receive element signal prior to being summed together, the result is that only the reflections from the 1 m wide azimuth portion arrive in phase, with all other ground returns being canceled out, or at least severely attenuated.

Now suppose the same process is done in sequence, rather than all at once. We start with an element at one end of the antenna, transmit a pulse, and receive the return using only this element. All the other elements are inactive. Both transmit and receive signals are modified by the phase compensation as before. The return sequence is stored in memory. Then we repeat this process with each separate antenna element in turn until we have saved all the 301 return sequences. Remember, these return sequences are complex numbers, with a magnitude and phase. Now if we sum all the complex results at the end, we must have the same result as if we did everything in parallel at once. Nothing else has changed; the situation on the ground is assumed static. This is a simplified version of the process the SAR radar performs. Imagine as the plane flies forward, the PRF is such that 301 pulses are transmitted and received along 301 m of flight path.

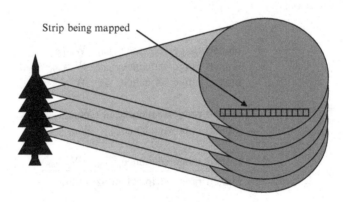

Strip being mapped

Figure 18.2

The radar could then effectively map the 1 m wide strip at right angles to the flight path. However, while this method solves the azimuth resolution problem, it is still not workable because only 1 m wide of strip ground is mapped perpendicular to the plane's flight path every 301 m, as shown in Figure 18.2.

18.4 SAR Processing

To go further, we need some conventions. Let us assign an index to each 1 m strip of land, oriented at right angles to the flight path, designated "n." At the PRF when the aircraft is physically aligned with $strip_n$, the real antenna receives a complex range $sequence_n$. This same index applies to the virtual or synthetic antenna element that is directly perpendicular to the 1 m strip. The next synthetic antenna element forward would be index $n + 1$, continuing on up to $n + 150$. The synthetic antenna element behind would be $n - 1$, extending to $n -150$. We have complex weighting factors of proper phase and amplitude for each index, W_{-150} through W_{150}. W_0 is always equal to 1.

To calculate the image for $strip_n$, we must start receiving range sequences at index $n - 150$ and sort into range bins according to arrival time. This is using the single real antenna with a wide beam angle (using 6-12° is typical). This continues for 300 more PRF intervals and results in 301 stored receive sequences in memory, indexed from -150 to $+150$. After PRF_{150}, we can start processing. The 301 stored, binned range sequences are multiplied or scaled by W_{-150} through W_{150}, respectively. Note that both the values in the binned range sequences and the weighting factors are complex. Each range bin is multiplied separately by the weighting factor. Then the 301 complex range sequence results can all be summed together across each range bin and result in a range binned sequence that has an azimuth of 1 m. The range bins correspond to the individual values across the elevation of the 1 m $strip_n$:

$$
\begin{array}{ll}
 & \text{binned range sequence}_{n\text{-}150} \cdot W_{-150} \\
+ & \text{binned range sequence}_{n\text{-}149} \cdot W_{-149} \\
+ & \text{binned range sequence}_{n\text{-}148} \cdot W_{-148} \\
\cdots & \\
+ & \text{binned range sequence}_{n+0} \cdot W_0 \\
+ & \text{binned range sequence}_{n+1} \cdot W_{+1} \\
\cdots & \\
+ & \underline{\text{binned range sequence}_{n+150} \cdot W_{+150}} \\
= & \text{binned range strip}_n \text{ (1 m azimuth)}
\end{array}
$$

For $strip_{n+1}$, we wait until PRF_{151}, and the plane has advanced 1 m on its flight path. We can then start processing again. We have just saved binned range $sequence_{151}$ and can discard binned range $sequence_{-150}$. This updated set of 301 binned range sequences is again multiplied by the weighting factors W_{-150} through W_{150}. In this case, there is an offset as follows:

$$\begin{array}{ll}
 & \text{binned range sequence}_{n-149} \cdot W_{-150} \\
+ & \text{binned range sequence}_{n-148} \cdot W_{-149} \\
+ & \text{binned range sequence}_{n-147} \cdot W_{-148} \\
\cdots & \\
+ & \text{binned range sequence}_{n+1} \cdot W_{0} \\
+ & \text{binned range sequence}_{n+2} \cdot W_{+1} \\
\cdots & \\
+ & \underline{\text{binned range sequence}_{n+151} \cdot W_{+150}} \\
= & \text{binned range strip}_{n+1} \text{ (1 meter azimuth)}
\end{array}$$

In this manner, we can compute each of the strips, one after the other, using a single broad beam antenna, but using a long synthetic array to achieve a narrow azimuth. We can make the synthetic antenna arbitrarily long, by using more PRF cycles, more memory, and higher processing rates.

The signal processing achieves the same lobe cancellation of signals coming from azimuths outside our desired 1 m ground strip as an actual 300 meter antenna would do. This processing technique is known as line-by-line processing.

18.5 SAR Doppler Processing

Another alternative method to perform SAR processing is to incorporate Doppler processing. Due to the use of the efficient FFT algorithm, this leads to a much lower level of computations than line-by-line processing.

Consider the patch of ground being illuminated by the radar pulse directed at right angles to the aircraft flight path. This patch may be 2000 m or more wide (azimuth direction), depending on the range and antenna beamwidth azimuth. At the midpoint, the Doppler frequency is exactly zero because the radar is moving parallel to this point and has no relative motion. At the two azimuth end points in the scan area, we have

positive Doppler frequency = sin(azimuth angle)·(aircraft velocity)/(wavelength)

negative Doppler frequency = − sin(azimuth angle)·(aircraft velocity)/(wavelength)

As an example, with a range of 10 km and a ground scan area of ±1000 m, this equates to an angle of ±5.71°. If the aircraft is flying at 250 m/s, this works out to ±829 Hz in for the 10 GHz radar band. This is shown in the Figure 18.3.

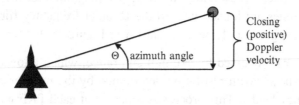

Figure 18.3

The Doppler frequency variation is not linear across the azimuth due to the sin(azimuth angle) in the equation. At small angles, $\sin(\theta) \approx \theta$, or approximately linear. As the angle increases, the effect becomes more nonlinear, until at 90°, the Doppler frequency asymptotically approaches the familiar (aircraft velocity/wavelength), or 8333 Hz. However, we want the Doppler frequency response to be completely linear across the azimuth range. We can compensate for this by using a phase correction multiplier, known as "focusing." The purpose is to make the Doppler frequency variation linear across the azimuth angle, rather than proportional to the sine of the angle. Once the frequency spacing per unit length on the ground is made linear, we can use Doppler filters with equally spaced main lobes along the frequency axis. This filtering is the familiar DFT, which can be implemented using the FFT algorithm. This is known as "SAR Doppler processing." The advantage of this approach is that the computational load is made much more manageable than the line-by-line processing technique, by virtue of the FFT algorithm efficiency.

As a side note, this Doppler linearity is an issue only for SAR radar. For conventional radar, the radar is aimed toward the horizon, and there is less variation due to the aspect angles (in this case, $\theta_{elavation}$ is close to 0°, although $\theta_{azimuth}$ can have significant variation) and the sensitivity requirements are much less than for SAR.

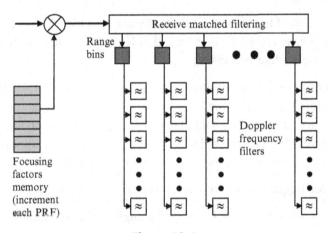

Figure 18.4

At each PRF, the return sequence is multiplied by a phase correction (focusing). Each range bin stores a complex value, representing the phase and magnitude of the return at that range. For each range, the values are loaded into all the Doppler frequency filters matching the azimuth angles for each ground element, as shown in Figure 18.4.

Each pulse has its return processed by azimuth and range, which allows separation over all locations in the radar beam, with resolution determined by the number of range bin and Doppler filter frequency banks. This process is repeated at each PRF with the next phase correction value, and the results are accumulated or integrated. Over the set of N PRFs equal to the number of N elements in the synthetic antenna, this is repeated.

After each set of N PRFs, the process repeats. The same point is measured N times, and complex values representing both magnitude and phase are integrated over the measurements for each point. In this architecture, the number of virtual elements in the synthetic array is equal to the number of Doppler filters, which can also be set equal to the number of range bins. This is also the number of times each point is measured, and the results integrated. However, each of the different measurements for a given point is done at a different azimuth angle.

In reality, these two methods provide equivalent results, although the processing steps are different. The first method is conceptually easier for most people to understand. The second method has the advantage of lower computational rate. An intuitive Figure 18.5 depicts the two different approaches.

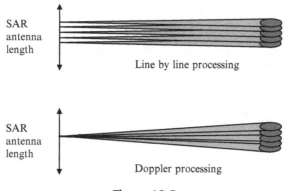

Figure 18.5

Another way to look at this is that in line-by-line, a new narrow beam at right angles to the flight line is synthetically created each PRF. With Doppler processing, many different azimuth beams are generated by each Doppler frequency bank during each PRF, and the returns from each beam are summed over multiple PRFs.

18.6 SAR Impairments

Several factors can degrade SAR performance. One of the most significant is the nonlinear flight path of an aircraft. We have seen how sensitive the phase alignments are to proper focusing, in fractions of the radar wavelength. Therefore, deviations in flight path away from the parallel line of the radar scan path must be determined and accounted for. This motion compensation can be done using inertial navigation equipment and by using GPS location and elevation measurements. Another consideration is side lobe return. When the side lobe return from the ground beneath the plane is integrated over a wide azimuth and elevation angles, this can become significant despite the low antenna gain at the side lobes. The design of the synthetic antenna, just like a real antenna, must take this factor into account. There are methods, similar to windowing in FIR filters, that can reduce side lobes, but at the expense of widening the main lobe and degrading resolution. Another issue is that the central assumption in SAR is that the scanned area is not in motion. If vehicles or other targets on the ground are in motion, they are not resolved correctly and are distorted in the images. Shadowing is another impairment. This occurs when a tall object shields another object from the radar's illumination, causing a block or blank spot in the range return. This issue becomes more prevalent when very shallow angles are used, which occurs when the aircraft is at a low altitude and scanning at long ranges. At high altitudes, such as satellite-mounted SAR, shadowing is much less of an issue.

CHAPTER 19

Introduction to Video Processing

Video signal processing is used throughout the broadcast industry, the surveillance industry, in many military applications, and is the basis of many consumer technologies. Up until the 1990s, nearly all video was in analog form. In a relatively short time span, nearly all video technologies have become digital. Virtually all video is now represented in digital form, and DSP techniques are used in nearly all video signal processing functions.

The picture element, or pixel, is used to represent each location in an image. Each picture element has several components, and each component is usually represented by a 10-bit value.

19.1 Color Spaces

There are several conventions, or "color spaces," used to construct pixels. The broadcast industry originally used black-and-white images, so a video signal contained only luminance (or brightness) information. Later, four-color information was added to provide for color television and movies. This was known as the chrominance information. The color space format associated with this is known as YCrCb, where Y is the luminance information, and Cr and Cb are the chrominance information. Cr tends to contain more reddish hue color information, whereas Cb tends to contain more bluish hue color information Each is usually represented as a 10-bit value. One advantage of this system is that image processing and video bandwidth can be reduced by separating the luminance and chrominance. The reason is that our eyes are much more sensitive to intensity, or brightness, than to color. So a higher resolution can be used for luminance and less resolution for chrominance. There are several formats used:

4:4:4 YCrCb	Each set of four pixels is composed of four Y (luminance) and four Cr and four Cb (chrominance) samples.
4:2:2 YCrCb	Each set of four pixels is composed of four Y (luminance) and two Cr and two Cb (chrominance) samples.
4:2:0 YCrCb	Each set of four pixels is composed of four Y (luminance) and one Cr and one Cb (chrominance) samples.

Most broadcast systems and video signals use the 4:2:2 YCrCb format, where the luminance is sampled at twice the rate of each Cr and Cb chrominance. Each pixel therefore requires an average of 20 bits to represent, as compared to 30 bits for 4:4:4 YCrCb.

Digital Signal Processing 101. DOI: 10.1016/B978-1-85617-921-8.00023-7

An alternate system was developed for computer systems and displays. There was no legacy black and white to maintain compatibility with, and transmission bandwidth was not a concern because the display is just a short cable connection to the computer. This system is known as the RGB format, for red/green/blue. Each pixel is composed of these three primary colors and requires 30 bits to represent. Most all televisions, flat screens, and monitors use RGB video, whereas nearly all broadcast signals use 4:2:2 YCrCb video.

These two color spaces (see Figures 19.1 and 19.2) can be mapped to each other as follows:

$$Y = 0.299 \cdot R + 0.587 \cdot G + 0.114 \cdot B$$

$$Cr = 0.498 \cdot R - 0.330 \cdot G + 0.498 \cdot B + 128$$

$$Cb = -0.168 \cdot R - 0.417 \cdot G - 0.081 \cdot B + 128$$

and

$$R = Y + 1.397 \cdot (Cr - 128)$$

$$G = Y - 0.711 \cdot (Cr - 128) - 0.343 \cdot (Cb - 128)$$

$$B = Y + 1.765 \cdot (Cb - 128)$$

Figure 19.1

Figure 19.2

There are other color schemes as well, such as CYMK, which is commonly used in printers, but we do not cover this scheme further here.

Different resolutions are used. Very common is the National Television System Committee (NTSC), also known as the SD or standard definition. It has a pixel resolution of 480 rows and 720 columns. This resolution forms a frame of video. In video jargon, each frame is composed of lines (480 rows) containing 720 pixels. The frame rate is approximately 30 frames (actually 29.97) per second, or fps.

19.2 Interlacing

Most NTSC SD broadcast video is interlaced. This was due to early technology in which cameras filmed at 30 frames per second, but this was not a sufficient update rate to prevent annoying flicker on television and movie theater screens. The solution was interlaced video where frames are updated at 60 frames per second, but only half of the lines are updated on each frame. On frame N, the odd lines are updated, and on frame N + 1 the even lines are updated and so forth. This is known as "odd and even field updating."

Interlaced video requires half the bandwidth to transmit as noninterlaced, or progressive, video at the same frame rate because only one half of each frame is updated at the 60 fps rate.

Modern cameras can record full images at 60 fps, although there are still many low-cost cameras that produce this interlaced video. Most monitors and flat-screen televisions usually display full, or progressive, video frames at 60 fps. When you see a 720p or 1080i designation on a flat screen, the "p" and "i" stand for progressive and interlaced, respectively. This has become less of an issue, as nearly all new products are now 1080p.

19.3 Deinterlacing

An interlaced video stream is usually converted to progressive for image processing, as well as to display on nearly all computer monitors. Deinterlacing must be viewed as interpolation because the result is twice the video bandwidth. There are several methods available for deinterlacing, which can result in different video qualities under different circumstances.

The two basic methods are known as "bob" and "weave." Bob is the simpler of the two. Each frame of interlaced video has only one half the lines. For example, the odd lines (1, 3, 5, ... 479) have pixels, and the even lines (2, 4, 6, ... 480) are blank. On the following frame, the even lines have pixels, but the odd lines are blank. The simplest bob interlacing is to just copy the pixels from the line above for blank even lines (copy line 1 to line 2) and copy the pixels from the line below for blank odd lines (copy line 2 to line 1). Another method would be to interpolate between the two adjacent lines to fill in a blank line. Both of these methods are shown in Figure 19.3.

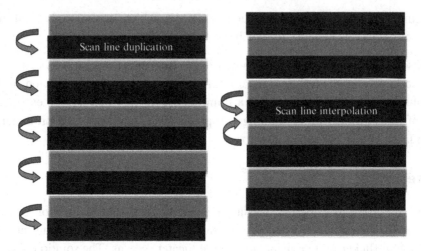

Figure 19.3

This method can cause blurring of images because the vertical resolution has been effectively halved.

Weave deinterlacing creates a full frame from the separate interlaced frames with odd and even lines. It then copies this frame twice to achieve the 60 fps rate. This method tends to work only if there is little change in the odd and even interlaced frames, meaning there is little motion in the video. Because the odd and even frame pixels belong to different instances in time (1/60th of a second difference), rapid motion can result in jagged edges in the images rather than smooth lines. This is shown in Figure 19.4.

Figure 19.4

Both of these methods have drawbacks. A better method, which requires more sophisticated video processing, is to use motion adaptive deinterlacing. Where there is motion on the image, the bob technique works better, and slight blurring is not easily seen. In still areas of the image, the weave method results in crisper images. A motion adaptive deinterlacer scans the whole image and detects areas of motion by comparing to previous frames. It uses the bob method in these areas of the frame and uses the weave method on the remaining areas of the frame. In this way, interlaced video can be converted to progressive with little loss of quality.

19.4 Image Resolution and Bandwidth

Early televisions used a cathode ray gun inside a fluorescent tube. The gun traversed the screen horizontally from left to right for each line. There was a brief horizontal blanking period, while the gun swung back to the left side of the screen, as indicated by the horizontal sync (HSYNC) signal. After 480 lines, the gun would be at the bottom-right corner of the screen. It would swing back to the top-left corner to begin a new frame, as indicated by the vertical sync (VSYNC) signal. This time period was the vertical blanking time.

Due to these blanking periods, the frame size is larger than the image size. For example, in SD definition, the actual image size viewed is 480 × 720 pixels. When blanking times are included, this is as if a 525 × 858 pixel image was sent, with the pixels blanked out, or zeroed, during the extra horizontal and vertical space. This legacy of allowing time for the cathode ray gun to return to the beginning of the video line or frame is still present in video standards today. In digital video and displays, these empty pixels can be filled with what is called ancillary data. This data could be display text at the bottom of the screen, audio information, data on the program currently being viewed, and so forth. The extra blanking times must be taken into account when determining video signal bandwidths.

Higher definition or resolution, or HD, video formats are now common. HD can refer to 720p, 1080i, or 1080p image resolutions. Many people are unaware, but much of the HD video content they receive from their cable or satellite provider is actually 720p video. The quality difference is small, and it requires half the transmission bandwidth to transmit. Most HD flat screens can display 1080p resolution, but this resolution video is often available only through DVRs or from other in-home media sources.

Table 19.1: Common video resolution and data rates

Image Size	Frame Size	Color Plane Format at 60 fps	Bps Transfer Rate
1080p × 1920	1125 × 2200	4:2:2 YCrCb	2200 × 1125 × 20 × 60 = 2.97 Gbps
1080i × 1920	1125 × 2200	4:2:2 YCrCb	2200 × 1125 × 20 × 60 × 0.5 = 1.485 Gbps
720p × 1280	750 × 1650	4:2:2 YCrCb	1650 × 750 × 20 × 60 = 1.485 Gbps
480i × 720	525 × 858	4:2:2 YCrCb	858 × 525 × 20 × 60 × 0.5 = 270 Mbps

19.5 Chroma Scaling

Chroma scaling is used to convert between the different YCrCb formats, which have various resolutions of color content. The most popular format is 4:2:2 YCrCb. In this format, each Y pixel has alternately a Cr or a Cb pixel associated with it, but not both. To convert to the RGB format, 4:4:4 YCrCb representation is needed, where each Y pixel has both a Cr and Cb pixel associated with it. This requires interpolation of the chroma pixels. In practice, this is often done by simple linear interpolation or by nearest neighbor interpolation. It can also be combined with the mapping to the RGB color space. Going the other direction is even simpler because the excess chroma pixels can be simply not computed during the RGB → 4:2:2 YCrCb conversion.

19.6 Image Scaling and Cropping

Image scaling is required to map to either a different resolution or to a different aspect ratio (row/column ratio). This process requires upscaling (interpolation) over two dimensions to go to a higher resolution, or downscaling (decimation) over two dimensions to go to a lower resolution. Several methods of increasing complexity and quality are available:

- Nearest neighbor (copy adjacent pixel)
- Bilinear (use 2 × 2 array of 4 pixels to compute new pixel)
- Bicubic (use 4 × 4 array of 16 pixels to compute new pixel)
- Polyphase (larger array of N × M pixels to compute new pixel)

When several lines of video are filtered vertically, the memory requirements increase because multiple lines of video must be stored to perform any computations across vertical pixels. The effects of increasing filtering when downscaling can be easily seen using a circular video pattern. Downscaling, like decimation, causes aliasing if high frequencies are not suitably filtered, as shown in Figure 19.5.

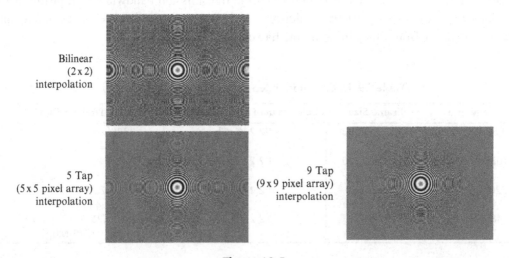

Bilinear
(2 x 2)
interpolation

5 Tap
(5 x 5 pixel array)
interpolation

9 Tap
(9 x 9 pixel array)
interpolation

Figure 19.5

Upscaling is far less sensitive than downscaling. Bicubic (using a 4 × 4 array of pixels) is sufficient. The effect of performing upscaling by using a smaller array or nearest neighbor is limited to slight blurriness, which is far less objectionable than aliasing.

Cropping is simply eliminating pixels, to allow an image to fit within the frame size. It does not introduce any visual artifacts.

19.7 Alpha Blending and Compositing

Alpha blending is the merging of multiple images. One image can be placed over the top of another image, as shown in Figure 19.6. This method is known as compositing, a method commonly used to implement "picture in picture" functionality.

The more general case is a blending or weighting of the pixels in each image. This process is controlled by the factor alpha (α). This blending is done on a pixel-by-pixel basis, for each color as shown:

$$\text{New pixel}_{red} = \alpha \cdot \text{pixel}_{red} \text{from image2} + (1 - \alpha) \cdot \text{pixel}_{red} \text{from image1}$$

$$\text{New pixel}_{green} = \alpha \cdot \text{pixel}_{green} \text{from image2} + (1 - \alpha) \cdot \text{pixel}_{green} \text{from image1}$$

$$\text{New pixel}_{blue} = \alpha \cdot \text{pixel}_{blue} \text{from image2} + (1 - \alpha) \cdot \text{pixel}_{blue} \text{from image1}$$

Figure 19.6

19.8 Video Compression

Video data is very large, due to two spatial dimensions, high resolution, and requires 60 fps. To store or transmit video, data compression technology is used. This capability is essential to allow services such as video on demand or streaming video to handheld wireless devices. Video compression is a lossy process; some information is lost. The goal is to achieve as much compression as possible while minimizing the data loss and restoring same perceptual quality when video is decompressed. This gaol is especially important in fast-motion video, such as during televised sports.

Video compression ratios depend both on the compression technology or standard used, as well as the video content itself. The newer video compression algorithms can deliver better quality, but at a price of very high computational requirements. Video compression processing is almost always done in hardware due to the computational rate, either in FPGAs or ASICs.

The most popular video compression algorithms are part of the MPEG4 standard. MPEG4 evolved from the earlier H.263. At the time of this writing, MPEG4 part 10, also known as MPEG4 AVC or H.264, is being widely adopted by the industry.

Most other video compression algorithms are variants of MPEG4. One example is QuickTime, developed by Apple Computer. Another is Windows Media Video, developed by Microsoft.

As you might guess, video compression is a very complex topic. It involves analysis across both spatial and temporal dimensions, and uses complex algorithms. In Figure 19.7, the two images depict comparative quality of an early version of MEPG4 and quality of the later MPEG4-10 in a video with a high degree of motion.

Figure 19.7

19.9 Video Interfaces

There are several common video interfaces used both in the broadcast industry and among consumer products. These are briefly described here:

- HDMI: Commonly used in high definition home theater systems. Often used to connect to interconnect flat screen televisions, Blu-ray video recorders and cable boxes. HDMI is a high quality shielded cable, to transmit uncompressed 1080p video in digital form.
- SDI: This is a broadcast industry standard, used to interconnect various professional equipment in broadcast studios and mobile video processing centers (like those big truck trailers seen at major sporting events). SDI stands for serial data interface, which is not very descriptive. It is an analog signal, modulated with digital information. This is usually connected using a coaxial cable. It is able to carry all the data rates listed in the table shown earlier and dynamically switch between them. Most FPGAs and broadcast ASICs can interface directly with SDI signals.
- DVI: This is a connection type common on newer computer monitors. It is a multipin connector carrying separated RGB digitized video information at the desired frame resolution.
- Analog RGB: This is an interface used to connect most computer monitors. It is the familiar multipin "sub-D" connector located on the back or side of all computers, used to connect to monitors or for laptops to connect to projectors for whole-room viewing. This interface carries separated RGB digitized video information at the desired frame resolution.
- CVBS: Standing for Composite Video Blanking and Sync, this is the basic yellow cable used to connect televisions, VCRs, and DVDs together. It carries an SD 4:2:2 YCrCb combined analog video signal on a low-cost coax "patch cable."
- S-Video: This is commonly used to connect consumer home theater equipment such as flat-panel televisions, VCRs, and DVDs together. It carries a 4:2:2 YCrCb signal in separate form over a single multipin connector, using a shielded cable. It is higher quality than CVBS.
- Component Video: This is also commonly used to connect consumer home theater equipment such as flat-panel televisions, VCRs, and DVDs together. It carries a 4:2:2 YCrCb analog signal in separate form over three coax patch cables. Often the connectors are labeled Y, PB, and PR. It is higher quality than S-video due to the separate cables.

CHAPTER 20

Implementation Using Digital Signal Processors

A digital signal processor is a special form of microprocessor that is optimized to perform DSP operations. The first DSP processors became available in 1982. They are designed to perform DSP functions in an efficient manner, using conventional software programming design and verification techniques. Due to the special features and parallelism added for DSP operations, conventional programming languages such as C often do not have the syntax to express the needed operations, although advanced compilers developed by the DSP processor manufacturers do try to interpret the programmer's intent and map to the DSP processor features in the most effective manner. For this reason, a vendor-specific language, known as assembly, is instead used to code the most DSP-intensive portions of the software. Assembly language is a proprietary set of instructions for a given processor. This is not as big a restriction as it may sound because the portion of the software code implemented in assembly is often 5–10% or less of the entire code base, and the DSP vendors themselves often provide the reference code for many common DSP algorithmic implementations. But writing in assembly code does require knowledge of the DSP processor hardware architecture and how to optimally write code that takes advantage of the DSP processor architecture features.

20.1 DSP Processor Architectural Enhancements

DSP processor enhancements, as compared to those of microprocessors, are generally divided into three categories:

- They enable maximum data bandwidth in and out of the DSP core engine.
- They efficiently implement the mathematical operations commonly required in DSP.
- They allow for the addition of multiple cores or hardware coprocessors to meet demands of the high computational rate of DSP applications.

20.1.1 Data I/O Bandwidth

All processors can be limited by data bandwidth in and out of the processor core or engine. However, bandwidth is especially a concern for DSP processors. This concern is due to the nature of the processing tasks. Many microcontrollers are primarily used to make

Digital Signal Processing 101. DOI: 10.1016/B978-1-85617-921-8.00024-9

decisions based on various inputs and to control various tasks in response. A flow chart or state diagram is often used to describe the required behavior. In contrast, a DSP processor typically is focused on processing or performing specific mathematical operations on continuous streaming input data, often in an assembly line fashion. Where decisions are made, they are typically based on some mathematical characteristic of the input data, which the DSP processor has determined after on-the-fly analysis of the data.

Data is transferred on buses, which can be 16, 32, 64, or even larger bit widths. Data buses are used to connect the core to various memories and to various peripheral units. The memories might be on-chip RAM or off-chip Flash or DRAM. Peripheral units could be serial ports, parallel ports, USP ports, Ethernet MACs, and so on. The DSP core is primarily interfacing with on-chip memory. On-chip memory is usually very fast and low latency. Data usually can be read or written in a single processor clock cycle. This is also known as Level 1, or L1, memory. Larger multi-cycle access memory, which can also be on-chip, is known as L2 memory. Much larger off-chip memory, which has even longer access times, is often referred to as L3 memory. Caches can also be used, but typically only for the instructions because DSP data is often read or written only once from memory.

DSP processors, at a minimum, need three input data buses. On any given DSP processor clock cycle, the core must be able to read the next instruction word and two data words. This is sometimes referred to as the "Harvard" architecture. The two data words are typically both the digitized signal being processed and a fixed coefficient. This would certainly apply to FIR filters as well as FFTs (where the coefficient is the complex exponential or twiddle factors). Often a fourth bus is employed to write back to the on-chip memory. Data buses can sometimes have additional flexibility, such as a 32-bit bus to be able to read or write 16-bit complex symbol pairs in a single cycle, by concatenating the quadrature component.

In addition, DSP processors often need to read or write data in a specific order to optimize the processing steps. Recall in our discussion of FFTs that the FFT data is either read in or written out in "bit-reversed" format. Or imagine performing digital filtering, where after processing one sample through all the filter coefficients, the next sample processing is required to read the same set of coefficients again. Or data may need to be read or written in various strides (reading every N^{th} sample) for decimation filters. To facilitate these functions, the DSP processors use hardware units known as data address generators (DAGs). A DSP processor typically has multiple DAGs (four or more) to be able to reconfigure DAGs during a given algorithm. A DAG can be programmed to support a circular buffer, for example. In this case, the DAG might read in a set of coefficients sequentially and, at the end, wrap around to start anew at the top of the coefficient memory location. This is known as circular addressing. Another typical function might be to read or write data with a programmable stride. Often, the I and Q values might be stored in an interleaved fashion. If a function needs to access the real and imaginary data separately, two DAGs might be programmed to access data at every other memory location, with the quadrature data DAG starting an offset relative to the real data DAG.

For other data transfers, such as from one type of memory to memory, or from a peripheral unit to memory, special hardware units known as direct memory access (DMA) controllers are used. The advantage is to allow the processor to preprogram these DMA controllers to move a block of data from one place to another, relieving the processor core from spending time and clock cycles in performing these functions itself. This capability is especially important in DSP processors because data is usually processed in chunks or blocks, and must be available to the DSP core when ready for processing. Typically, the DSP core processes one block of data while the DMA engine is moving the next block of data to a designated location in L1 memory.

A simple example might be an input signal coming through an ADC. The ADC samples a signal and converts it into a digitized form at a specific sampling rate. The ADC data could be transferred to the DSP through several possible serial or parallel interface methods. The DMA engine might be configured to move these samples sequentially into a specific processor memory space and interrupt the processor when a specific number of samples have been received. This eliminates the need for the DSP core to perform individual read operations, and be interrupted only when a complete block of data is ready for processing. Like the DAGs, the DMA engines in DSP processors are often capable of supporting different strides and offsets. DMA engines may have the flexibility to support multidimensional data structures, which might be composed of frames, which in turn are composed of slots, which are composed of symbols, which are composed of real and quadrature data words.

20.1.2 Core Processing

Due to the high processing rates required and the need to complete processing in a deterministic time interval (known as real-time processing), DSP processors usually execute all instructions in a single clock cycle. This is similar to RISC processors, but in the case of DSP processors, the instructions can be quite complex. In fact, an architecture known as very long instruction word (VLIW) has become popular with several DSP processor vendors because it allows for complex instructions to be easily defined and executed in a single cycle, using very parallel core hardware architectures. It can also be adapted to work with a variable-length instruction word, depending on each instruction's complexity. The most common instructions would tend to be the shortest, reducing code memory size.

The most fundamental DSP operation is to multiply two operands and add the product to an accumulator. This is known as a multiply accumulate (MAC). Naturally, the DSP core contains one or more multiply-accumulator circuits. In fact, this is so important that DSP processors are often rated in either millions of instructions per second (MIPs) or, more commonly, in millions of multiply accumulates per second (MMACs). The most common precision size is a 16×16-bit multiplier, feeding a 40-bit accumulator. This means that there are 8 extra bits in the accumulator, and 2^8, or 256, products can be summed into the accumulator before any accumulator overflow could occur. Some DSP processors also have a provision for 32×32-bit multiplication, often by combining several 16-bit multipliers. Architectural support for single-cycle complex multiplication using multiple multipliers in the core is also advantageous.

Larger multipliers can be built from smaller multipliers, allowing a flexible multiplier size. As shown here, a single 32×32 multiplier can be built from four 16×16 multipliers, using several adders:

$$\text{Inputs}: \quad A[31:0] \quad B[31:0]$$

$$\text{Output}: \quad R[63:0]$$

The four 16×16 multipliers produce the following products:

$$A[15:0] * B[15:0] \quad = P1[31:0]$$
$$A[31:16] * B[15:0] \quad = P1[47:16] \text{ must shift left by 16}$$
$$A[15:0] * B[31:15] \quad = P1[47:16] \text{ must shift left by 16}$$
$$A[31:16] * B[31:16] = P1[63:32] \text{ must shift left by 32}$$

The sum of these four products forms the final 32×32 result, R[63:0].

A 16×16 complex multiplier is also implemented using four 16×16 multipliers:

$$(A + jB) \cdot (C + jD)$$
$$= \quad A \cdot C + jB \cdot C + A \cdot jD + jB \cdot jD$$
$$= \quad AC + jBC + jAD - BD$$
$$= \quad (AC - BD) + j(BC + AD)$$

The real part of the complex product is the difference between the two multiplier products. The imaginary part of the complex product is the sum of the two multiplier products. Notice that four multiply operations are required.

Barrel shifting, rounding, and saturation circuits and other shift instructions are also important for DSP processing. A simple example could be implementation of FIR filters. The input data is multiplied by all the coefficients, each output requiring N cycles for an N tap FIR filter. As the result is accumulated over the N cycles, the result grows. Assuming a 16-bit DSP and a 40-bit accumulator, the filter output is 40 bits. The accumulator can often be saved as the LSW (least significant word, 16 bits), MSW (most significant word, 16 bits), and overflow (8 bits). Often, only the MSW result is saved back to memory as a 16-bit result.

This requires shifting the result as needed to align with MSW boundaries, rounding the result using the LSW content, and performing saturation in the event the result exceeds this representation. All these functions should be able to be performed in a single cycle as the 16-bit result is saved back to memory, and so provisions for this must be provided in the vendor-specific assembly instructions.

Other necessary functions are supported for Boolean operations such as AND, OR, and EXOR. These are normally performed across a 16- or 32-bit word. In addition, bit field packing, concatenation, rotation, and extraction are often highly useful for various error correction and cryptography applications, and are included in a DSP processor instruction set.

DSP processors spend most of their cycles in tight loops. Therefore, they have provisions for what is called "zero overhead looping." In a conventional processor, the code must often decrement a counter and test the result at the end of the loop to determine whether to exit or jump to top of the code loop again. In DSP processors, this capability is built into the hardware, so no extra cycles need to be used in this testing or jumping back to the top.

Due to the need to respond quickly in external inputs, DSP processors have resources to facilitate low-latency interrupt service routines (ISRs). Features such as interrupt shadow registers, vectored ISRs (rather than use of single global ISR locations), nested interrupt capabilities, and other operations allow for a DSP processor to react quickly, prioritize, and perform necessary processing from a number of different interrupt sources, including the DMA controller engines.

DSP processors may also have application-specialized instructions. For example, vector instructions might allow simultaneous processing of real and quadrature portions of a symbol. Often, there is an ADD, COMPARE & SELECT instruction to allow efficient implementation of the Viterbi algorithm.

20.1.3 Multiple Cores or Hardware Coprocessors

DSP vendors have also responded to the high computational demands of LTE wireless baseband processing in particular by building hardware circuits to perform specialized LTE algorithmic tasks, thereby off-loading the DSP cores. Otherwise, the DSP cores would be unable to keep up with the required LTE baseband processing. Due to the large LTE base station market for DSP processors and the well-defined requirements in the LTE standard, it has been economically feasible to optimize DSP processor products specifically for this application. Functions like Turbo error correction and OFDMA FFT processing can be implemented in hardware coprocessors, with the data flow being managed by one or more of the multiple DSP cores contained in a single device.

20.2 Scalability

In many applications, the trend is for increasing DSP computational requirements. For example, use of high definition in video processing, 4 G LTE systems, advanced radar and sonar, and some new error correction algorithms requires increasingly high rates of DSP.

In the 1990s, DSPs evolved by increasing clock rates, increasing amounts of L1 memory, and some architectural improvements. In the next decade, further improvements such as more onboard multiply accumulate circuits per core, better C compilers, addition of Ethernet MAC, and other high-bandwidth interfaces were included. However, silicon process technology improvements were unable to deliver the same improvements in clock circuit speeds. Clock rates have topped out at about 1.2 GHz in the highest performance DSPs. To deliver more processing power, DSP processor vendors have responded by adding more DSP cores per chip, currently up to six independent cores.

Having more processor cores is effective when many different and independent DSP tasks can be partitioned and run simultaneously. It is a poor solution when the tasks are interrelated, or individual tasks have very high computational requirements. It is also not especially scalable. As the number of cores increases, partitioning the tasks across the cores and managing the intercore communications become increasingly difficult. Several DSP processor "startup" or venture-funded companies have developed products with dozens or hundreds of cores, but these products were not significantly adopted by industry.

20.3 Floating Point

Nearly all DSP processors are fixed-point processors. This is due to the larger floating-point circuit requirements, higher power consumption, and lower performance compared to fixed point. For the vast majority of DSP applications, fixed-point arithmetic is suitable. However, this does require care on the part of the designer to ensure the dynamic range of the signal is mapped into the limited fixed-point precision.

Several DSP processors do offer floating-point DSP products, although at much lower performance levels than the fixed-point products. One popular application for floating-point DSP processors is high-fidelity audio processing. This is due to relatively low processing rates, high dynamic range requirements, and extensive use of IIR filters, which can have stability issues when implemented in fixed point.

High-performance floating point is more commonly implemented on Pentium-type processors, graphics processors, or some specialty floating-point processors. Usage tends to be on military applications, such as radar back-end processing, or high-performance computing, such as for research purposes (an example might be climate simulations).

20.4 Design Methodology

The design methodology used with DSP processors is very similar to that used on other types of processors. The software-based approach offers the optimal flexibility to build and debug complex algorithms. Due to the serial nature of the software flow, the implementation and debugging are simplified, all variables in memory are accessible, and the flow tends to be more "natural" to the designer's thought process.

In summary, the DSP processor is a specialized processing engine that is reconfigured each clock cycle for many different functions, mostly DSP related, others more control or protocol oriented. Resources such as processor core registers, internal and external memory, DMA engines, and I/O peripherals are shared by all tasks, often referred to as "threads." This creates ample opportunities for the design or modification of one task to interact with another, often in unexpected or nonobvious ways. In addition, most DSP algorithms must run in real time, so unanticipated delays or latencies can cause system failures. Some of the challenges of DSP programming include

- Mixture of C or high-level language subroutines with assembly language subroutines
- Possible pipeline restrictions of some assembly instructions
- Nonuniform assumptions regarding processor resources by multiple engineers simultaneously developing and integrating disparate functions
- Ensuring interrupts completely restore the processor state upon completion
- Blocking of a critical interrupt by another interrupt or by an uninterruptible process
- Undetected corruption or noninitialization of pointers
- Properly initializing and disabling circular buffering addressing modes
- Preventing memory leaks, the gradual consumption of available volatile memory due to failure of a thread to release all memory when finished
- Dependency of DSP routines on specific memory arrangements of variables
- Unexpected memory rearrangement by optimizing linkers and compilers
- Use of special "DSP mode" instruction options in core
- Conflicts or unexpected latencies of data transfer peripherals and memory when using DMA controllers
- Corrupted stack or semaphores
- Subroutine execution times dependent on input data or configuration
- Pipeline restrictions of some assembly instructions

20.5 Managing Resources

Interaction between different tasks or threads can cause intermittent and sometimes hard-to-detect problems in all processor architectures. Microprocessor, DSP, and operating system (OS) vendors have attempted to address these problems with different levels of protection or isolation of one task or "thread" from each other. An operating system can be used to manage access processor resources, such as allowable execution time, memory, or common peripheral resources. However, there tends to be an inherent compromise between processing efficiency and the level of protection offered by the OS. In DSPs, where processing efficiency and deterministic latency are often critical, the result is usually minimal or no level of real-time operating system (RTOS) isolation between tasks. Each task or thread often requires unrestricted access to many processor resources in order to run.

The diagram in Figure 20.1 helps illustrate how complex a DSP processor system is. All these functions exist to service the DSP processor core, which can execute only one instruction at a time. But generally, only a subset of the hardware in a DSP processor is needed at any given time because the hardware must be designed to support every instruction. This is the inherent inefficiency in any processor architecture, compared to a custom hardware implementation. The penalty of the flexibility of the DSP processor is this hardware inefficiency.

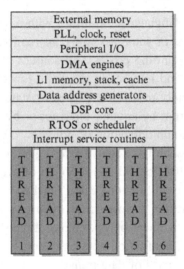

Figure 20.1

20.6 Ecosystem

A significant factor in choosing a silicon platform for DSP applications is the available tool and intellectual property (IP). Different DSP processor vendors have various amounts of DSP software IP available. Smaller, independent DSP IP companies also license software modules for many functions on popular DSP families. This can be a major factor in the choice of DSP processor. For example, someone wishing to build media gateways needs to implement voice compression and decompression for a variety of industry-standard voice codecs, or vocoders, as well as echo cancellation capability. These algorithms are available for several DSP processor families, eliminating the need for a proprietary development. Similar IP is available for image compression used in the video surveillance industry, motor control algorithms used in industrial applications, code for facsimile and modem standards, and many other applications.

In addition, major DSP processor manufacturers have sophisticated and robust development tools, which include compilers, assemblers, linkers, and debuggers. They also often supply RTOS, Ethernet TCIP stacks, and many other commonly used device drivers and IP stacks. In addition, many third-party companies also supply RTOS products and various IP stacks for specific DSP processor architectures.

CHAPTER 21

Implementation Using FPGAs

Field Programmable Gate Arrays (FPGAs) can be used to perform DSP. From their origins as custom logic and interface functions, FPGAs have grown to include nearly every function that can be implemented digitally and are able to interface to nearly any other circuit at nearly any data rate desired. From a DSP point of view, FPGAs are often used to interface to ADCs and DACs, as well as backplanes. Wherever a DSP processor is used, the FPGA often is used in the system for interfacing as well.

FPGA fabric is composed of configurable logic elements and programmable routing, which allow any digital function to be built, including multipliers, adders, accumulators, shifters, registers, and any other functions that might be found in a DSP processor. Distributed memory in different-size arrays or blocks is also integrated within FPGAs. In addition, to optimize FPGAs for DSP applications, special DSP blocks are available in most FPGA devices. Logic-intensive functions like multipliers can be hardened, which, besides leaving more programmable logic available for other functions, allows the multipliers to run much faster and consume less power. There are other circuits that can be hardened in addition to the multipliers. Typically, these circuits are collected into DSP blocks, which are architected to implement common DSP operations efficiently. This topic is discussed in more detail later in this chapter.

FPGA devices can be very small, from a few thousand logic elements, or very large, up to one million or more logic elements. The number of hardened multipliers can vary from a few dozen to thousands in a single device. Because of the sheer number of available circuits, FPGAs are much more powerful DSP platforms than any DSP processor. A high-end DSP might be able to perform perhaps 10 giga multiply accumulates (GMACs) per second, whereas a high-end FPGA could perform in excess of 1000 GMACs per second. This is an increase of about 100-fold, or two orders of magnitude.

With this disparity in processing power, you might wonder why DSP processors are so commonly used. DSP processors first became available in 1982, whereas FPGAs suitable for DSP applications were not available until the mid-1990s. Many applications have a legacy of using DSP processors. More importantly, the design

Digital Signal Processing 101. DOI: 10.1016/B978-1-85617-921-8.00025-0

methodology of the two technologies is completely different. This point is very important and helps explain why either a DSP processor or an FPGA is more suitable for a given application.

21.1 FPGA Design Methodology

FPGAs are fundamentally hardware devices, even though they are programmable. Unlike a processor, an FPGA is not controlled by scheduling the operations of a processing engine, but by configuring the hardware itself to perform the necessary operations for a particular design.

A DSP processor core is a reconfigurable engine, which on a cycle-by-cycle basis is configured to support a specific operation. Aside from the parallelism built into the core (e.g., parallel circuits to fetch the next instruction plus two data operands, while multiplying and accumulating previous operands), each algorithm is fundamentally implemented in a serial fashion. Processing rates are primarily determined by the DSP processor clock rate. The software designer of a processor-based system is limited by the available instructions of a given processor, which in turn depends on the processor architecture. Operations not supported in the processor hardware can still be implemented in software, but usually inefficiently. For example, a 16-bit division operation in most DSP processors is implemented iteratively over at least 16 clock cycles because there is no single-cycle division instruction.

By contrast, in an FPGA, the hardware is normally configured to support a predetermined DSP data path function. The entire application is normally implemented by building a separate hardware circuit for each operation and passing the data through in a process not unlike the assembly line used in factories. Just like on an assembly line, the distribution of operations to separate circuits results in a dramatic speedup in throughput because many operations are performed in parallel in a pipelined or step-by-step fashion. The DSP processor analogy would be a single worker to build the entire product, step by step. Throughput is increased by adding more workers (DSP processor cores).

Due to the distributed nature of memory in an FPGA, the needed data can always be quickly accessed from memory by each different circuit. The composite memory I/O bandwidth in the distributed FPGA memory far exceeds that of a DSP processor bandwidth, despite the multiple memory buses of the DSP processor.

An FPGA circuit can be made adaptable, often by using registers to set the operational mode. For example, an FFT hardware circuit can be built with a register-configurable number of points (although often limited to 2^N, where N is an integer), or an FIR filter can have coefficients stored in a memory block or registers, which can be updated during operation. Despite this, an FPGA simply does not have the same flexibility as a DSP processor. The processor can execute any valid instruction on the next cycle, whereas the flexibility needed in an FPGA hardware data path must be anticipated and provisioned for by the designer.

DSP processors, while having much smaller memory bandwidth, do have the capability to access any location across the entire memory space from cycle to cycle. FPGAs, while having much higher memory bandwidth due to inherent parallelism, can access only the data available in each local memory block, which is normally determined by the data flow through the data path as anticipated by the designer. Again, the FPGA achieves a massive increase in bandwidth, but at the expense of runtime flexibility.

21.2 DSP Processor or FPGA Choice

Given their differences, FPGAs and DSP processors tend to be used in applications where their respective merits can provide the optimum implementation platform. DSP processors tend to be more suitable when the application is very complex or requires many configurations, and the processing rates are low enough to allow for DSP processor implementation. Complex algorithms are often easier to implement and debug in a software-based flow. Some applications may require very data-dependent processing algorithms, where different DSP operations may be needed, as determined by the input data. In an FPGA, this would require several alternative hardware circuits, which can be complex and inefficient to implement in hardware.

In some organizations, a large DSP processor legacy code base used in product development and an experienced DSP processor programming team on staff can make use of an FPGA unattractive, despite its greater processing capabilities.

FPGA usage is prevalent when high rates of data throughput and processing are required. In terms of GMACs, a single FPGA can replace a whole board full of DSPs. Many times, the application processing rate dictates the implementation method. For example, high-definition image-compression algorithms such as H.264 require use of a hardware-based solution such as an FPGA, whereas standard (low) definition image compression may be able to be processed in real time by a suitable DSP processor. Radar systems often have extremely high rates of data throughput and may use multiple FPGAs chained together.

Even in many higher performance DSP systems, it is common to find both FPGAs and DSP processors in use. In such systems, FPGAs are generally used for preprocessing, filtering, and performing the main DSP data path operations. DSP processors may be used for back-end processing once the data rate has been reduced, or for more complex portions of an algorithm. An example is an adaptive filter. The FPGA may perform the actual filtering, but the DSP processor could be used to process any feedback information used by an adaptive algorithm to update the filter coefficients.

Small FPGAs can be useful coprocessors for DSP processors. When standard but high MMAC functions like digital filters and FFTs are offloaded, the DSP processor MIPs are freed up for more value-added functions. Implementing this type of setup is also often very feasible because in many DSP processor-based systems, an FPGA is already present in many cases to interface between the data convertors and the DSP processor, or between the backplane and the DSP processor.

21.3 Design Methodology Considerations

Designing with FPGAs is inherently more difficult than designing with a DSP processor. This is due to the high degree of choice in an FPGA device. The designer is able to create any hardware circuit desired, create any size data bus, configure the data flow as needed, plus synthesize internal microprocessors out of logic, and support nearly all possible serial and parallel external interfaces. This capability results in many available degrees of design freedom. The standard design entry method is known as Hardware Description Language (HDL). Two variants are commonly used: Verilog and VHDL. One HDL issue is compile time. On processors, new software updates can be compiled in a matter of seconds. With FPGAs, compile times can take minutes or hours. This is again due to many degrees of freedom. The structures described in the HDL code must be synthesized, a process not unlike that of compiling on a processor. The HDL code is broken down and synthesized into many small logic functions. It then must be mapped to FPGA hardware resources, and all interconnections made using the FPGA routing resources. The FPGA vendor-provided tools, known as place and route, perform this function while simultaneously ensuring the connection and logic delays still allow the design to operate at the clock rate specified by the designer. Verification is also much more arduous than on DSP processors, due to the need to verify not only logical operation, but the timing of all circuits and routed connections. Complex test benches that simulate as many possible states of the design in operation as possible are used for verification.

Surprisingly, FPGA designs can be more robust than DSP processor code implementations, despite the increased design effort. This is fundamentally due to the independence of the different tasks because of the inherent parallelism of the FPGA. Each task has separate hardware, including memory structures. This tends to limit the number of unexpected interactions that can occur when all functions share the same hardware and memory. The separation in an FPGA is shown in Figure 21.1, which is in contrast to the structure used in processors shown in the preceding chapter.

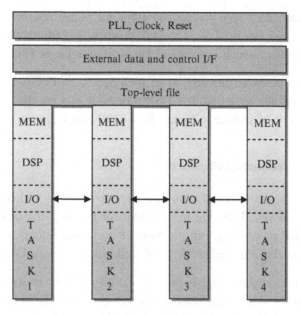

Figure 21.1

Additionally, FPGA verification methodologies are patterned after that of application specific integrated circuit (ASIC) design flows. With ASICs, millions of dollars and dozens of man years are often spent on each chip, so verification is a major function in the design flow. Bug fixes such as a simple recompile and software download are not possible. This has led to a huge investment in design verification tools by the EDA industry that can also be used to verify FPGA designs.

21.4 Dedicated DSP Circuit Blocks in FPGAs

Initially, it may appear that the hardened DSP block circuits in FPGAs should be designed to have the same capabilities as DSP processor cores. After all, most of the DSP algorithms to be implemented are the same. However, there are two major differences. First, in addition to DSP blocks, the FPGA also contains large amounts of programmable logic, which can also be used to implement DSP functions. So, in the FPGA's DSP block circuits, unlike a DSP processor, not all functional capabilities need to be implemented. Second, the large FPGAs may contain several thousand DSP block circuits. Therefore, the size of these blocks may significantly affect the silicon area and power consumption of the FPGA. Intelligent trade-offs need to be made on what should be hardened in the DSP block and what should be left in soft or programmable logic. FPGA vendors need to partition their silicon areas between DSP circuits, memory blocks, configurable logic, routing resources, high-speed I/O circuits, and hardened interface protocol circuits. By eliminating seldom-used functions within the FPGA DSP block circuit, FPGA vendors can reduce DSP block area and instead allow for higher numbers of DSP block circuits for the silicon area that is devoted to DSP processing. Optimizing this area also reduces power consumption, which is a critical factor in ever larger FPGA devices.

There are a number of potential features to be included in an FPGA-hardened DSP circuit. These features are normally included in a DSP processor and are discussed one by one, with a view on the merit of including them into a DSP block circuit.

21.4.1 Adjustable Precision Multipliers

In most FPGA products, the standard multiplier size is 18×18, rather than the 16×16 size used in DSP processors. This size is sufficient for the majority of applications. Some FPGAs, like some DSP processors, can use four 18×18 multipliers to implement a single 36×36 multiplier, or else a complex 18×18 multiplier within a DSP block.

There is also a growing need for higher precision multipliers in many applications, at least in some parts of the data path. Most data converters are 12–16 bits, and 18-bit multipliers are sufficient. Yet in some applications, the processing gain in decimation filters, or by algorithmically combining several correlated receivers' data (MIMO), can lead to an increase in data path precision. In FFTs, data precision naturally increases on only one side of the multiplier because the data path precision grows with each successive butterfly stage, but the complex exponential coefficients remain of fixed precision. Also, some DSP algorithms and applications just naturally require higher precision to meet their performance requirements. This can be accommodated by using higher-precision multipliers. However, use of a 25×25-bit multiplier requires twice the multiplier area of an 18×18 multiplier. This is a large penalty on applications requiring only 18-bit multipliers or less. And use of 36×36 multipliers built from 18×18s is also very expensive, particularly if the needed precision is only a few bits more than 18. Fortunately, FPGA multiplier circuits can be designed with more flexibility than DSP processors because they are not intrinsically tied to the data bus widths and instructions of a DSP processer. With 9×9 multipliers as building blocks, FPGA DSP blocks could support 9×9, 18×18, 27×27, 36×36, and 54×54 multiplier sizes with a roughly proportionate increase in DSP block resources as multiplier precision increases. The DSP circuits could be configured to allow either a high count of lower precision multipliers or a lower count of high-precision multipliers, which would allow system designers to design with the precision they need, rather than try to tailor to the limitations of the hardware device. When this approach is used, only the applications actually using this additional precision need to allocate more multiplier resources.

This capability is useful in a high percentage of DSP applications and cannot be efficiently implemented in soft logic. Therefore, this is a good capability to include in the DSP block structure of an FPGA.

21.4.2 Accumulators

Accumulators are integral to many DSP operations. They are necessary when using a single multiplier to calculate in series a number of multiplication and addition operations. The accumulator circuit can also be reused in another mode, the distributed adder circuit, described next. Accumulators are essentially large adders with feedback on one operand. They can be implemented in logic but then may not run as fast as the hard multiplier. Since they are used in a high percentage of DSP applications, this is a good capability to include in the DSP block of an FPGA. The size of the accumulator depends on the common multiplier precisions used. Normally, at least 8 extra bits should be used above the product size of the multiplier, to allow at least 256 product accumulations. An 18×18 multiplier size needs a 44 or more bits accumulator, a 27×27 multiplier size needs a 62 or more bits accumulator, and a 36×36 multiplier size needs an 80 or more bits accumulator. However, the larger accumulator size is a penalty on all DSP blocks, even when smaller multiplier sizes are used, although a much smaller penalty than an oversized multiplier. A good compromise would be to accommodate at least 18×18, 18×27, and 27×27 multiplier sizes because they are likely to be used frequently. This would lead to an accumulator size of about 62 bits or so.

21.4.3 Postadder (Subtracter) and Distributed Adder

Postadders are used to construct larger adders from smaller adders, to perform the sum/subtraction in complex multiplications, and to perform sums of products used in FIR filters. A postadder also should be able to perform subtraction if desired. This is an excellent function to include in a DSP block. An example of a postadder used in an FIR filter is depicted in Figure 21.2.

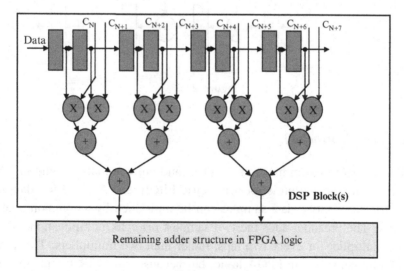

Figure 21.2

An alternate FIR filter architecture is known as the systolic architecture. It uses a distributed output adder, and therefore, no matter how large the FIR filter, no programmable logic-based adder circuits are required. This can allow for more efficient and higher Fmax (clock frequency) FIR filter implementations. The only penalty is an increase in filter latency and a slightly more complex sequencing of input data. This is a small price to pay for the benefits. With the inclusion of an accumulator or postadder, there is little extra cost to add this useful feature to an FPGA DSP block. A vertical cascade path is needed between DSP blocks, which are normally placed in columns in an FPGA. This fixed path is of insignificant cost, and could also be used to build large multiplier circuits that span several DSP blocks, such as 54 × 54 or complex 27 × 27. For these reasons, the cascade path is an excellent function to include in a DSP block.

An example of a postadder used in a systolic FIR filter is depicted in Figure 21.3. The postadder is connected as a distributed adder cascaded from block to block. Note the additional registers in the input data chain to compensate for the postadder register delay stages. This example performs the same algorithmic function as the diagram in Figure 2.12, although with greater latency (clock delay).

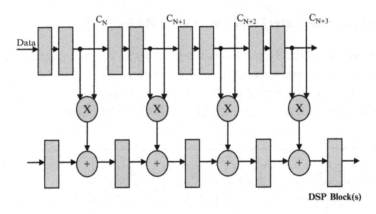

Figure 21.3

21.4.4 Preadder (Subtracter)

Preadders are primarily used in hardware circuits and not commonly found in DSP processors. The main application is for symmetric FIR filters. As the filter data is shifted across the coefficient set, two data samples can be multiplied by a common coefficient due to the symmetry. The preadder adds the two samples *prior* to multiplication, which allows the use of one multiplier for every two taps, rather than two multipliers. This preadder function can be implemented in FPGA logic, but because most FIR filters are symmetric and FIR filters are the most common DSP application, this is a reasonable feature to include in a DSP block. Configured as a subtracter, this block can also be used to perform a "sum of absolute differences," a less commonly used function. Also, although not discussed here, in some cases a preadder can also be used to more efficiently implement complex multiplication architectures.

The preadder is difficult to implement with DSP processors because normally a single multiplier is used, with the input data being in a circular buffer. The input data is therefore not organized in a way to easily take advantage of a preadder.

Two symmetric eight-tap filters are diagramed in Figure 21.4. The filter taps are [C_0, C_1, C_2, C_3, C_3, C_2, C_1, C_0]. A preadder is used on the left in a conventional FIR filter and on the right in a systolic FIR filter implementation.

The extra input data registers in the systolic diagram used to align data flow are shown in the datapath at the top of Figure 21.4.

The input data needs to "wrap around" to utilize the preadder, which makes it difficult to implement in DSP processors. In both conventional and systolic FIR filter architectures, it can reduce multiplier usage by approximately one half. This is a valuable feature to incorporate into a DSP block.

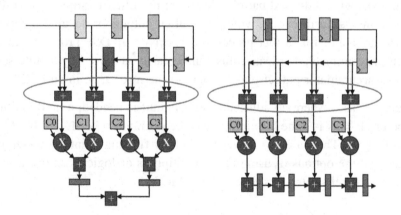

Figure 21.4

21.4.5 Coefficient Storage

Coefficients are required for most DSP operations, including FIR filters and FFTs. These coefficients are normally stored in distributed memory blocks external to the DSP block. This allows for large numbers of coefficients to be used, and for easy updating in the case of adaptive filters, for example. However, since most FIR filters in a hardware implementation are built using a parallel, or at least partially parallel, structure, the number of coefficients used per multiplier is often fairly small. For these cases, it may be advantageous to allow internal coefficient storage, dynamically selectable on each clock cycle, *inside* the DSP block. The advantage, besides an obvious savings of FPGA memory resources, is reduced routing congestion and logic in the region near the DSP block inputs. The logic and routing in this area of the design is usually in high demand, due to the input data routing and need to build variable-length shift registers to provide proper input data flow, especially for FIR filters incorporating preadders, interpolation, and/or decimation structures. Use of internal coefficient registers also eliminates the possibility of being unable to run at the required circuit Fmax due to routing congestion. There is a trade-off in number of coefficients, DSP block area required, and percentage of applications where the internal coefficient bank contains enough coefficients to accommodate the FIR filter. However, on the balance, this seems to be a valuable feature to consider including in an FPGA DSP block.

21.4.6 Barrel Shifter

Shifters are necessary in DSP processors and are an integral part of a DSP processor instruction set. Shifting is often possible in parallel with many other DSP processor operations. Surprisingly, dedicated barrel shifters are not often required in a DSP block structure. The principal justification, which is to align outputs or decimal points for 16- or 32-bit memory storage or for further processing, is driven by DSP processor architecture limitations. In programmable hardware, this can be easily achieved by simply selecting which subset of accumulator output pins to route to the next stage in an FPGA.

For the remaining cases where an adjustable shift circuit is needed, a hard multiplier circuit can be used. This is done by multiplying the operand to be shifted by 2^N, which achieves a shift by N. This can also be very useful for floating-point normalizations. The low usage and the options of using either multipliers or logic as shift registers argue against hardening barrel shifter functions in DSP blocks.

21.4.7 Rounding and Saturation

Rounding and saturation functions are necessary in DSP processors to reduce the output data width after multiple accumulate or distributed adder circuits. This step must be implemented in hardware; there is no practical way to implement it in software. In FPGAs, only a small fraction of the DSP blocks tend to need rounding and saturation. The reason is partly that, in a distributed adder filter, only the last stage tends to be rounded or saturated. There are also a number of different rounding methods used by system designers in different applications. Hardening this relatively silicon-costly function across thousands of DSP blocks makes little sense when it can be efficiently implemented in logic in the instances where it is required, with the flexibility to use any preferred rounding method. The one exception is to offer "biased rounding," the simplest rounding method (other than truncation), which can be implemented almost free in the existing accumulator/adder circuit.

21.4.8 ALU and Boolean Operations

ALU and Boolean operations include ADD, SUBTRACT, AND, OR, and EXOR type functions. Normally, use of these functions is not associated with multiply accumulate operations. Therefore, it is wasteful to use a DSP block to implement these simple ALU operations, since the great majority of any DSP block area is used to build multiply accumulate circuits. This is better implemented in logic wherever needed and can be efficiently built using soft logic. Extra dedicated circuits for these functions are not recommended for DSP block inclusion, due to the additional complexity and circuitry.

21.4.9 Specialty Operations

Some DSP processors have special instructions for bit-reversed addressing for FFTs, modes for Viterbi FEC processing, or instructions to perform bit-by-bit correlation necessary in CDMA applications. FPGAs need not support these functions in a hard DSP block; they are not often used in a high percentage of DSP blocks and can be easily implemented in logic.

21.4.10 Tco and Fmax

Most DSP blocks in high-performance FPGAs can be clocked in excess of 500 MHz. However, delays in logic and routing limit design of the Fmax to 400 MHz or less in nearly all designs of any size. Further increases in DSP block Fmax are not especially useful, due to this mismatch in logic and routing speeds, and very high Fmax circuits require a disproportionate amount of current, even when not being clocked at high rates.

What is more useful is to minimize the Tco, or settling time relative to output clock. Using Tco reduces the routing delay contributions in the DSP block to route to the next stage of logic. This helps reduce routing delays at the DSP block output, where routing congestion is often high and, as a result, increases system clock frequency. To achieve lower Tco requires eliminating unnecessary circuits after the last register stage in the DSP block. This is another reason to remove some of the previously listed features that are of questionable value in the DSP block.

21.5 Floating Point in FPGAs

The vast majority of DSP applications utilize fixed point. However, some high-performance systems require the dynamic range and precision of floating point. Some examples would be high-resolution radar systems, STAP radar processing, and MIMO channel calculations in LTE wireless systems. Typical functions implemented in floating point could be FFTs, matrix multiplication, matrix inversion, trigonometric operators, and others.

Single-precision floating point uses a 23-bit mantissa, plus a sign bit. This requires a minimum of 24 × 24-bit multiplier for floating-point multiplication. Floating point also requires normalization and denormalization. This requires the capability to shift a 24-bit operand within a 48-bit field. This can be implemented as a barrel shifter in logic or, for faster circuit operation, using a minimum of 24 × 24-bit multiplier and 48-bit accumulator.

In processor-based implementations, normalization and denormalization are performed at each floating-point calculation. Recent FPGA design techniques and IP are available that allow for high-performance floating-point data paths to be implemented. This typically requires extra mantissa bit width. The extra mantissa width eliminates the need for normalization and denormalization at each floating-point calculation, which can significantly reduce routing and logic resources, allowing higher performance FPGA implementation. This technique can utilize larger multipliers, such as 27 × 27 or 36 × 36-bit precision.

The capability to efficiently implement both high-performance fixed-point and floating-point DSP data paths allows designers the flexibility to use whatever precision is required at different points in their systems. Even in applications where floating point is required, the majority of the processing can still be implemented in various precision fixed-point circuits. FPGAs can provide the flexibility to build both types of implementations.

21.6 Ecosystem

As with DSP processors, FPGA vendors also have ecosystems of tools and IP, provided both by the FPGA manufacturers and third-party companies. Each FPGA vendor provides comprehensive software tools to enable the various steps of synthesis, fitting, timing analysis, and many other functions. Due to the complexity and breadth of this software, there can be considerable variation in the feature set, robustness, and ease of use between FPGA vendors. Third-party companies also supply tools for the FPGA development process, especially for verification purposes.

The major FPGA vendors also supply comprehensive collections of IP modules, much of which are for DSP applications. Interestingly, FPGA and third-party vendors also offer several synthesizable "soft" microprocessor architectures, which are built from the configurable logic within the FPGA. These microprocessors can deliver quite respectable performance and run a number of operating systems, including Linux. Apart from the relative merits of the FPGA silicon device itself, often the tools, IP, and reference designs can play a significant factor in choosing an FPGA platform for DSP and other applications.

21.7 Future Trends

It is generally perilous to forecast the future, but some trends are apparent and worth commenting on. In terms of their processing capabilities, DSP processors are increasing rather slowly, due to inabilities to significantly increase clock rates. Scaling by providing more cores or more multipliers per core increases design flow complexity, and tends to be effective only on applications where the tasks can be partitioned among several cores and dependency between tasks is limited. More cores or more parallels within the cores also tend to make the design process more complex, partially negating the simplicity advantage DSP processors enjoy in design methodology.

FPGAs, in contrast, are increasing in processing capabilities with each succeeding new product family. While FPGA circuit clock rates are increasing slowly, as with DSP processors, the logic density of the FPGA devices is nearly doubling every 2 years. This capability merely increases the existing parallelism of FPGAs and does not fundamentally alter the design complexity or methodology. This is allowing ever more complex and higher computational rate algorithms to be implemented, thereby increasing system capabilities.

Furthermore, FPGA design methodologies are also improving. Alternatives to HDL, the main flow used by FPGA designers at the time of this writing, are now available. These alternatives can provide higher productivity by allowing designers to work at a higher level of abstraction. One design flow is known as "model-based" design flow. It uses the popular tool Simulink by The MathWorks© to describe the design. The test bench can also be implemented in Simulink, or alternatively by using another popular MathWorks© tool called MATLAB. Use of these tools can allow very rapid design changes and verification, compared to traditional HDL design flow. Recent innovations in this flow by one FPGA vender can also generate a very high Fmax FPGA circuit performance, similar to the best HDL-generated designs.

Another design methodology that multiple vendors are promoting is known as "C to gates" design. In this flow, a software description of the hardware is used. This design flow can lead to higher productivity and easier design reuse. A second advantage is a potential huge new pool of FPGA designers because there is at least an order of magnitude more C programmers than HDL-knowledgeable engineers.

Many approaches to the "C to gates" design are using a variant of the C language known as System C, which has provisions not only to describe the normal data processing steps and decision flow, but also to describe what is to be implemented in parallel and what in serial (traditional) software flow. Products now exist to support this methodology, but at present, adoption is limited due to both tool immaturity and just natural conservatism. This is a revolutionary design methodology change that will require time rather than evolutionary change in the FPGA design flow. Still, this is very likely the direction of the future.

In addition, many evolutionary changes are occurring in the HDL design flow. Continuous reductions are occurring in the FPGA compile times. Partial recompile, which eases small updates into the design, is now available from some FPGA vendors, as is partial reconfiguration, which allows reprogramming of very small selected sections of the FPGA circuitry while the FPGA is in operation. New debugging capabilities are becoming available. Many of these changes are expected to narrow the design effort gap between DSP processors and FPGAs for DSP systems and applications.

Q Format Shift with Fractional Multiplication

Consider normal multiplication as follows:

$$
\begin{array}{r}
104 \\
\times \quad 512 \\
\hline
53{,}248
\end{array}
$$

In this example, there is no decimal point consideration because both numbers are integers.

Now look at the same number, but with a different decimal point arrangement:

$$
\begin{array}{r}
10.4 \\
\times \quad 5.12 \\
\hline
53.248
\end{array}
$$

How do we know where to place the decimal point if doing this multiplication by hand? One way is to use a sanity check such as $5 \times 10 = 50$, so the answer 53.248 obviously has the correct decimal point. But the way most of us learned in grade school was that the product should have the same number of digits to the right of the decimal point as the two multiplicands combined have to the right of their decimal point. The first number, 10.4, has one digit to the right of the decimal point, and the second number, 5.12, has two digits to the right of the decimal point. So the product, 53,248, should have three digits to the right of the decimal point, giving 53.248.

This same concept works for fractional binary multiplication. For example, one Q1.15 number multiplied by another Q1.15 number gives a 32-bit result. But where is the decimal point? Based on the preceding rule, the product should have $15 + 15 = 30$ bits to the right of the decimal point. Our result is then a Q2.30 number. If we use only the upper or most significant 16 bits, as is common in many implementations, the result is a Q2.14 number. By performing a single left shift after the multiplication, we get a Q1.15 number using the upper 16 bits, which is the same format as our input data (or a Q1.31 number if we chose to keep all 32 bits). This is the reason that many DSPs have fractional multiply instructions or modes that incorporate an extra left shift of the result. In FPGA implementation, the user can easily choose the output format of the multiplier and take this effect into account. The example from Chapter 1 on numerical representation is repeated here:

$$0x4000 \qquad value = {}^1\!/_2 \text{ in Q.15}$$
$$\times \; 0x200 \qquad value = {}^1\!/_4 \text{ in Q.15}$$
$$\overline{}$$
$$0x0800\;0000 \quad value = {}^1\!/_{16} \text{ in Q31}$$

After left shifting by one, we get

$$0x1000\;0000 \quad value = {}^1\!/_8 \text{ in Q31} \text{ — the correct result!}$$

If we use only the top 16-bit word from the multiplier output, after the left shift we get

$$0x1000 \quad value = {}^1\!/_8 \text{ in Q15} \text{ — again, the correct result!}$$

In DSP applications, we often perform multiply accumulate operations, where the result of many multiplies are summed to a common result (remember the FIR filter structure).

With all these additions, how can the sum be represented without overflow occurring? How does this affect our decimal point?

The solution is to make the gain of the filter equal to one. As explained in Chapter 5 on FIR filters, the sum of all the filter coefficients is the gain of the filter. By scaling the filter coefficients (dividing each coefficient by the sum of the coefficients), we can easily set the gain to unity. Then there is no possibility of overflow, and the decimal point is not e affected. The frequency response is also unchanged because each coefficient is being scaled equally. For example, suppose our filter coefficients are {1,3,5,3,1}. The sum of these coefficients is 13. Therefore, our scaled coefficients are {$^1\!/13$, $^3\!/13$, $^5\!/13$, $^3\!/13$, $^1\!/13$}. This allows us to represent the coefficients in fractional format, since they are all between −1 and +1, and guarantees that the decimal point will not be altered in the signal as it passes through our filter. We still need the left shift at the output of the multiplier, however. Alternately, in FPGA, we could easily apply the left shift to the accumulated sum, instead of at each multiplier output.

Evaluation of FIR Design Error Minimization

Let's minimize the expression

$$\text{Error} = \xi = \int_{-\pi}^{\pi} |\xi(\omega)|^2 d\omega$$

where

$\xi(\omega) = D(\omega) - H(\omega)$

$D(\omega) = $ desired frequency response

$$H(\omega) = \sum_{i=-\infty \text{ to } \infty} C_i e^{-j\omega k} = \text{actual frequency response}$$

We start by setting the derivative of the error ξ with respect to the filter coefficients (the one parameter that we have control over and want to optimize) equal to zero:

$$d|\xi|/dC_i = 0$$
$$d|\xi|/dC_i = \int_{-\pi}^{\pi} (d/dC_i)\{[D(\omega) - H(\omega)] \cdot [D(\omega) - H(\omega)]^*\} d\omega = 0$$

We then use the chain rule to differentiate each part:

$$d|\xi|/dC_i = \int_{-\pi}^{\pi} (d/dC_i)\{[D(\omega) - H(\omega)]\} \cdot [D(\omega) - H(\omega)]^* d\omega +$$
$$\int_{-\pi}^{\pi} [D(\omega) - H(\omega)] \cdot (d/dC_i)\{[D(\omega) - H(\omega)]\}^* d\omega$$

The derivative is now a sum of two integrals. Since they are complex conjugates, the sum can be zero only when both terms are zero. So we can consider only one of the two terms and evaluate when it is equal to zero:

$$\int_{-\pi}^{\pi} [D(\omega) - H(\omega)] \cdot (d/dC_i)\{[D(\omega) - H(\omega)]\}^* d\omega = 0$$

Now let's try to simplify the term $(d/d\,C_i) \{[\, D(\omega) - H(\omega)\,]\}^*$:

$$(d/dC_i)\{|D(\omega) - H(\omega)|\}^* = (d/dC_i)\{D(\omega)^* - (d/dC_i)\{H(\omega)^*\}$$

Notice that $(d/dC_i)\{ D(\omega)^* \} = 0$, since $D(\omega)^*$ does not depend on C_i. $D(\omega)$ is our desired response, and not dependent on C_i.

Recall that $H(\omega) = \sum_{k=-\infty \text{ to } \infty} C_k \, e^{-j\omega k}$ We have to replace the index "i" with "k," since we are already using "i" in this discussion as the coefficient index. Therefore,

$$H(\omega)^* = \sum_{k=-\infty \text{ to } \infty} C_k * e^{+j\omega k}$$

We want to simplify

$$H(\omega)^* = (d/dC_i) \left\{ \sum_{k=-\infty \text{ to } \infty} C_k * e^{+j\omega k} \right\}$$

Each coefficient C_i is independent of the others—for example, $(dC_1/dC_2) = 0$. Only when $i = k$ is there a nonzero result. Therefore, we can remove the summation and get

$$H(\omega)^* = (d/dC_i) \left\{ \sum_{k=-\infty \text{ to } \infty} C_k * e^{+j\omega k} \right\} = (d/dC_i)\{C_i * e^{+j\omega i}\} = e^{+j\omega i}$$

Now let's go back and substitute the simplified result:

$$(d/dC_i) \, \{|D(\omega) - H(\omega)|\}^* = (d/dC_i)\{D(\omega)^*\} - (d/dC_i)\{H(\omega)^*\} = e^{+j\omega i}$$

Now we are getting close. Let's go back to the original integral and substitute again:

$$\int_{-\pi}^{\pi}[D(\omega) - H(\omega)] \cdot (d/dC_i)\{[D(\omega) - H(\omega)]\} * d\omega = \int_{-\pi}^{\pi}[D(\omega) - H(\omega)] \cdot e^{+j\omega i} d\omega$$

Since we are trying to find the solution when the integral equals zero, we can equate the two terms within

$$\int_{-\pi}^{\pi} D(\omega) \cdot e^{+j\omega i} d\omega = \int_{-\pi}^{\pi} H(\omega) \cdot e^{+j\omega i} d\omega$$

Now let's substitute for $H(\omega)$ again. We get

$$\int_{-\pi}^{\pi} \sum_{k=-\infty \text{ to } \infty} C_k e^{-j\omega k} \cdot e^{+j\omega i} d\omega = \int_{-\pi}^{\pi} \sum_{k=-\infty \text{ to } \infty} C_k e^{-j\omega(k-i)} d\omega$$

Our goal is to get C_k by itself on the left side, since that is what we are trying to solve for.

We can rearrange the order of summation and integration, since both are linear operations:

$$\int_{-\pi}^{\pi} \sum_{k=-\infty \text{ to } \infty} C_k e^{-j\omega(k-i)} d\omega = \sum_{k=-\infty \text{ to } \infty} C_k \cdot \int_{-\pi}^{\pi} e^{-j\omega(k-i)} d\omega$$

Now let's evaluate the integral

$$\int_{-\pi}^{\pi} e^{-j\omega(k-i)} d\omega$$

There are two cases, when k = i, and when k ≠ i:

$$k = i \quad \int_{-\pi}^{\pi} e^{-j\omega(k-i)} d\omega = \int_{-\pi}^{\pi} e^{-j\omega 0} d\omega = \int_{-\pi}^{\pi} 1 d\omega = 2\pi$$

$$k \neq i \quad \int_{-\pi}^{\pi} e^{-j\omega(k-i)} d\omega = -j\omega(k-i) \cdot [e^{-j\omega(k-i)} |_{-\pi}^{\pi}] = -j\omega(k-i) \cdot [e^{-j\pi(k-i)} - e^{j\pi(k-i)}]$$

Let's consider two possibilities: the number m = k − i is even, or it is odd:

m is even [$e^{-j\pi m} - e^{j\pi m}$] = 1 − 1 = 0, since both of these terms are at the point (1 + 0j) on the unit circle.

m is odd [$e^{-j\pi m} - e^{j\pi m}$] = −1 − (−1) = 0, since both of these terms are at the point (−1 + 0j) on the unit circle.

So we get a very simple result:

$$\int_{-\pi}^{\pi} H(\omega) \cdot e^{+j\omega i} d\omega = \sum_{k=-\infty \text{ to } \infty} C_k \int_{-\pi}^{\pi} e^{-j\omega(k-i)} d\omega = C_k \cdot 2\pi$$

If we substitute this into

$$\int_{-\pi}^{\pi} D(\omega) \cdot e^{+j\omega i} d\omega = \int_{-\pi}^{\pi} H(\omega) \cdot e^{+j\omega i} d\omega$$

we finally arrive at the desired result:

$$C_k = (1/2\pi) \cdot \int_{-\pi}^{\pi} D(\omega) \cdot e^{+j\omega i} d\omega$$

The kth coefficient is found by multiplying the desired frequency response by $e^{+j\omega i}$ and integrating over the interval 2π. Each coefficient is computed independently. This method gives the lowest value of error ξ, as defined in Chapter 5 on FIR filters.

Laplace Transform

The Laplace transform is used for continuous or analog signals. Let's use x(t) and y(t) to represent the input and output signals to an analog filter, respectively. Rather than use delays, as digital filters use, we use the differentiator instead. The input and output of the analog filter have the following relationship:

$$y(t) = \sum_{i=0 \text{ to } N} A_i \cdot d^i x(t)/dt^i + \sum_{i=0 \text{ to } M} B_i \cdot d^i y(t)/dt^i$$

The first term is a sum of coefficient-weighted derivatives of the input, which is analogous to the FIR filter being a sum of weighted delayed inputs. The second term is a coefficient-weighted sum of the derivatives of the output, which implies feedback.

These functions can be difficult to evaluate and characterize. Smart people long ago figured out a way to map things into the s-domain using the Laplace transform, where the math becomes much simpler. This is an introduction to that technique.

We are going to use the exponential function again to determine the response of this filter. In this case, we use the function e^{st}, where s is a complex variable.

Notice this property of our chosen input, which we use to develop the s-transform:

$$de^{st}/dt = s \cdot e^{st} \text{ and } d^i e^{st}/dt^i = s^i \cdot e^{st}$$

If we apply this to our analog filter equation, we get

$$y(t) = \sum_{i=0 \text{ to } N} A_i \cdot s^i \, x(t) + \sum_{i=0 \text{ to } M} B_i \cdot s^i \, y(t)$$

We can therefore construct a Laplace-transform-based relationship of the filter response.

We want an equation with form y(t) = func(x(t)). After doing a bit of rearranging, we get

$$y(t) = x(t) \cdot \left\{ \left(\sum_{i=0 \text{ to } N} A_i \cdot s^i \right) \Big/ \left(1 - \sum_{i=0 \text{ to } M} B_i \cdot s^i \right) \right\}$$

where the bracketed term is the s-transform function, denoted as H(s):

$$H(s) = \left(\sum_{i=0 \text{ to } N} A_i \cdot s^i \right) \Bigg/ \left(1 - \sum_{i=0 \text{ to } M} B_i \cdot s^i \right)$$

The roots of the numerator give the zero locations of the Laplace transform, and the roots of the denominator give the pole locations of the s-transform. These zeros and poles characterize the filter frequency response.

Similar to digital filters, analog filters can also be defined by their impulse response. In the digital domain, an impulse is defined as a single sample with value = 1. For the analog world, this is not as simple. The answer is to define a function called the delta dirac, known as $\delta(t)$. It is a strange function. It has infinite height and zero width, but the area under its integral is equal to 1. It is essentially a spike located at zero along the number line. This is the continuous signal equivalent to a digital impulse. We can define the impulse response of an analog filter as follows:

$$\text{Impulse response} = h(t) = y(t)|_{x(t)=\delta(t)}$$

So the output is the impulse response when the input is $\delta(t)$.

Now recall that s is a complex number. We can evaluate the Laplace transform and determine its zeros (values of where the transform numerator goes to zero) and its poles (values of s where the transform denominator goes to zero), and plot these locations on the complex number plane. In this context, the complex number plane is known as the s-plane.

Next, let's go through a simple example to clarify:

$$y(t) = x(t) + dx(t)/dt - 2 \cdot (dy(t)/dt)$$
$$y(t) = x(t) \cdot \{(1+s)/(1+2s)\}$$

We can see that there is a single zero at $s = -1$, and a single pole at $s = -1/2$. The frequency response is evaluated by setting $s = j\omega$ and evaluating for $-\infty < \omega < \infty$.

The filter's response is determined by the pole and zero locations (see Figure C.1). Use of the Laplace transform facilitates the design of analog filters.

We have mentioned that filters with feedback can be unstable. Let's examine a simple case to see when this is true:

$$y(t) = A_0 \cdot x(t) + \cdot dy(t)/dt$$
$$y(t) = A_0 \cdot x(t) + B_1 \cdot s \, y(t)$$
$$H(s) = A_0/(1 - B_1 \cdot s),$$

which has a single pole at $s = 1/B_1$.

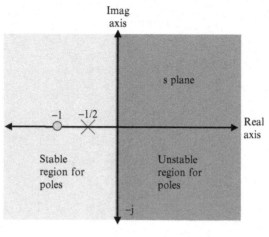

Figure C.1

To determine stability, we compute the impulse response of the filter. Because of the simplicity of this example, the response of filter with $x(t) = \delta(t)$ can be computed directly from the differential equation. We find that

$$h(t) = (A_0/B_1) \cdot e^{(t/B_1)} \text{ for } t \geq 0 \text{ and } h(t) = 0_j \text{ for } t < 0$$

If B_1 is negative, the exponential will decay over time—this is considered a stable response. But if B_1 is positive, the impulse response will grow to infinity over time—this is an unstable response. This leads to the following general rule with analog filters:

All poles must be on left side of the s-plane.

There is no restriction on the location of zeros.

Z-Transform

The z-transform is used for sampled signals. It is analogous to the s-transform used with continuous or analog signals. Let's use x_k and y_k to represent the input and output signals to a digital filter, respectively. Rather than use differentiators, as analog filters use, we use the delay function instead. The input and output of the IIR (recursive) digital filter have the following relationship:

$$y_k = \sum_{i=0 \text{ to } N-1} C_i\, x_{k-i} + \sum_{i=1 \text{ to } N-1} D_i\, y_{k-i}$$

The first term is a sum of coefficient-weighted delayed versions of the input (this is the familiar FIR filter expression). The second term is a coefficient-weighted sum of the delayed versions of the output, which implies feedback.

The feedback part prevents us from using our earlier method of FIR design analysis, described in Appendix B. But again, some smart people figured out a way to map things into the z-domain, where the math becomes simpler. This is an introduction to that technique.

The z-transform is very useful. It can be used to determine the frequency response of an IIR filter. It can also be used to find the coefficients of the IIR filter from the zeros' and poles' locations, or to do the inverse, to find the poles and zeros of the IIR filter from the coefficients.

The z-transform uses a special exponential function z^{-i}, where z is a complex variable.

The definition of the z-transform of any given sampled time sequence x_i is

$$X(z) = \sum_{i=-\infty \text{ to } \infty} x_i\, z^{-i}$$

We can characterize a digital filter by its z-transform, $H(z)$. We can show that

$$Y(z) = H(z){\cdot}X(z) \text{ or } H(z) = Y(z)/X(z)$$

where

    $Y(z)$ is the z-transform of the output sampled time sequence y_k.
    $X(z)$ is the z-transform of the input sampled time sequence x_k.

Let us consider a digital filter, with an impulse response of h_i. The output of this filter is given by

$$y_k = \sum_{i=-\infty \text{ to } \infty} h_i \, x_{k-i}$$

where x_k is the input sequence (we developed this in Chapter 5 of FIR filters).

Let's take the z-transform of both x_k and y_k to see if we can determine $H(z)$:

$$X(z) = \sum_{k=-\infty \text{ to } \infty} x_k \, z^{-k}$$

$$Y(z) = \sum_{k=-\infty \text{ to } \infty} y_k \, z^{-k} = \sum_{k=-\infty \text{ to } \infty} \left\{ \sum_{i=-\infty \text{ to } \infty} h_i \, x_{k-i} \right\} z^{-k}$$

Now we interchange the order of summation:

$$Y(z) = \sum_{i=-\infty \text{ to } \infty} \left\{ \sum_{k=-\infty \text{ to } \infty} h_i \, x_{k-i} \, z^{-k} \right\}$$

Next, we remove h_i from the inner summation, as it is independent of k:

$$Y(z) = \sum_{i=-\infty \text{ to } \infty} h_i \left\{ \sum_{k=-\infty \text{ to } \infty} x_{k-i} \, z^{-k} \right\}$$

Next, we factor z^{-k} into 2 terms:

$$Y(z) = \sum_{i=-\infty \text{ to } \infty} h_i \left\{ \sum_{k=-\infty \text{ to } \infty} x_{k-i} \, z^{-(k-i)} \, z^{-i} \right\}$$

Then we remove z^{-i} from the inner summation, as it is independent of k:

$$Y(z) = \sum_{i=-\infty \text{ to } \infty} h_i \, z^{-i} \left\{ \sum_{k=-\infty \text{ to } \infty} x_{k-i} \, z^{-(k-i)} \right\}$$

The bracket term in the preceding equation is simply $X(z)$.

$$H(z) = \sum_{i=-\infty \text{ to } \infty} h_i \, z^{-i} \text{ is the } z-\text{transform of impulse response } h_i.$$

This demonstrates that $Y(z) = H(z) \cdot X(z)$.

Let's look at this scenario a different way. The filter's response has a z-transform, which is equal to the z-transform of the output sequence y_k divided by the z-transform of the input sequence x_k, denoted $H(z)$.

By definition, y_k equals the impulse response when we set x_k equal to an impulse. The z-transform of the impulse function is equal to 1, so the z-transform of the impulse response of the filter is simply the z-transform of the output:

$$H(z) = Y(z)/X(z) = \sum_{i=-\infty \text{ to } \infty} h_i\, z^{-i}$$

where h_i is defined as the impulse response of the filter.

This result sounds familiar to frequency response $H(\omega)$, and indeed, the two are related. The frequency response is contained within the z-transform. The frequency response can be found from the z-transform by replacing z with $e^{j\omega}$, or when z is evaluated around the unit circle on the complex z-plane:

$$H(\omega) = \sum_{i=-\infty \text{ to } \infty} h_i\, e^{-j\omega i}$$

With an IIR filter, computing the impulse response sequence h_i can often be difficult. The usefulness of the z-transform is that it can be used to compute the frequency response $H(\omega)$ without first computing the impulse response h_i. The z-transform can be found easily if either the coefficients of the digital filter are available, or the poles and zeros of the digital filter are available. Then we take the expression $H(z)$, replace z with $e^{j\omega}$, and then evaluate the frequency response over the interval $-\pi < \omega < \pi$.

$H(z)$ is simply a function of z. Our goal is to express $H(z)$ as a function of two polynomials in z, in the form of

$$H(z) = A(z)/B(z)$$

With a bit of algebra, we can then rearrange the polynomials into the following form:

$$H(z) = \left(\sum_{i=0 \text{ to } N} C_i \cdot z^{-i} \right) \Big/ \left(1 - \sum_{i=1 \text{ to } M} D_i \cdot z^{-i} \right)$$

In this form, the coefficients are available by inspection. Or, if you have the coefficients, you can easily write the z-transform of the filter.

The z-transform can also be arranged in pole – zero format. Remember, by definition, the zeros of the z-transform are the values of z where $H(z) = 0$ (the roots of the numerator) and the poles of the z-transform are the values of z where $H(z) = \infty$ (the roots of the denominator):

$$H(z) = \left[\prod_{i=0 \text{ to } M} (z - \text{zero}_i) \right] \Big/ \left[\prod_{i=0 \text{ to } N} (z - \text{pole}_i) \right]$$

To summarize, if you are given the poles and zeros, you can immediately construct the z-transform using the preceding template. Although it may take a bit of algebra, the z-transform can be rewritten in summation form to find the coefficients. And evaluating the z-transform in either form over the complex unit circle gives the frequency response of the filter. This is done by substituting $e^{j\omega}$ for z and computing over the interval $-\pi < \omega < \pi$:

$$H(z) = \left[\prod_{i=0 \text{ to } M} (z - zero_i) \right] \Big/ \left[\prod_{i=0 \text{ to } N} (z - pole_i) \right] =$$

$$H(z) = \left(\sum_{i=0 \text{ to } N} C_i \cdot z^{-i} \right) \Big/ \left(1 - \sum_{i=0 \text{ to } M} D_i \cdot z^{-i} \right) \text{ and}$$

$$\text{Frequency response } H(\omega) = H(z = e^{j\omega})|_{\omega=-\pi}^{\omega=\pi}$$

As with the s-plane, we can show that pole location on the z-plane is restricted for stability reasons. All poles within the z-plane must lie within the unit circle. Zeros have no restriction on the z-plane.

Binary Field Arithmetic

Binary field arithmetic is used in coding. The basic rules are very simple. Addition is as follows:

$$0 + 0 = 0$$
$$0 + 1 = 1$$
$$1 + 0 = 1$$
$$1 + 1 = 0$$

This also corresponds to the exclusive OR or XOR operator in digital logic. Subtraction results are identical to addition; there is no distinction between the two operations.

Multiplication is as follows:

$$0 \cdot 0 = 0$$
$$0 \cdot 1 = 0$$
$$1 \cdot 0 = 0$$
$$1 \cdot 1 = 1$$

Division is not defined or allowed.

Linearity means that if two input codewords are input to a system, X_1 and X_2, the output result is the same as if the sum of two codewords $X_1 + X_2$ were input instead. A more mathematical way to express this idea is

$$Y_1 = f(X_1)$$

$$Y_2 = f(X_2)$$

For a linear system, we use

$$Y_2 + Y_1 = f(X_2 X_1)$$

Index

Note: Page numbers followed by *f* indicates figures; *t* indicates tables.

A

Aliasing, 22, 23, 26, 27*f*, 36, 63–65, 67, 98, 117, 118, 121, 198, 198*f*, 205, 205*f*, 209, 228, 229
Alpha blending, 229
Analog-to-digital convertor (ADC), 21, 25–26, 27, 28, 29, 37, 63, 65, 65*f*, 117*f*, 118, 119–121, 122, 235
Assembly language, 146–147, 233, 239

B

Bi-linear transform, 76–78, 79, 80
Bit-reversed addressing, 112, 253
Bob, 225, 226–227

C

Cartesian coordinates, 10, 11
Chrominance, 223
Code division multiple access (CDMA), 94, 150, 151, 152–155, 159–162, 163–165, 166–167, 169, 170, 171, 177–178, 180, 181, 182–183, 185*t*
Coding gain, 30, 125–126, 141
Common public radio interface (CPRI), 189–190
Complementary cumulative distribution function (CCDF), 185, 186
Complex conjugate, 15, 49, 200, 214, 259
Complex exponential, 15, 16–18, 31–32, 34, 35, 38, 45, 46, 47, 57, 99*t*, 101–102, 104–105, 106, 108, 111–112, 115, 119, 173, 174, 176, 177, 234, 248
Complex number, 9, 10, 12, 13, 14, 15, 16, 17–18, 19, 31–32, 49, 97, 217, 264
Convolutional encoding, 131–134, 147
Crest factor reduction (CFR), 180, 185–188, 189
Cyclic prefix, 177–180, 181
Cyclic redundancy check (CRC), 141–142

D

Decibels, 29, 30, 125–126, 183, 184, 186, 187
Decimation, 63–67, 71, 75, 112, 119–120, 146, 224–225, 228, 248, 251–252
Deinterlacing, 225–227
Digital downconversion, 114–115, 117, 119, 120, 204
Digital predistortion (DPD), 185, 188–189
Digital-to-analog converters (DAC), 29, 30, 91, 114–115, 116, 122–123, 127, 162, 180, 203, 243
Digital upconversion, 91, 113, 114–116, 123, 203
Discrete Fourier transforms (DFT), 97, 98–107, 108, 109, 173, 175, 220
Doppler ambiguities, 205–206
Doppler shift, 93, 149, 182, 183, 201, 202, 203

E

Error vector magnitude (EVM), 93–94, 185, 186–187, 188
Euler equation, 16, 17, 51
Exponent, 1, 7–8, 7*t*, 98–99, 101, 102

F

Fast Fourier transform (FFT), 97, 98–99, 105, 105*t*, 106–112, 146, 175, 182, 183, 200, 204, 219, 220, 234–235, 237, 244–245, 246, 248, 251–252, 253–254
Field programmable gate array (FPGA), 6, 7–8, 52, 66–67, 113, 230, 231, 243–246, 247–254, 255, 257–258
Fixed point representation, 2–7, 7–8
Floating point representation, 7–8
Forward error correction (FEC), 125, 253
Fractional representation, 1–2, 4–7, 5*t*

Frequency division duplex (FDD), 143, 145, 170, 180–181
Frequency modulation, 145–146, 197

G

Global system for mobile communications (GSM), 149, 166, 167, 169, 178, 185, 185t, 187

I

Infinite impulse response (IIR), 44, 73–76, 77–78, 79, 80, 238, 267, 269
Interlacing, 225
Intermediate frequency (IF) subsampling, 118–123
Interpolation, 63, 67–70, 71–72, 75, 91, 180, 225, 228, 251–252
Intersymbol interference (ISI), 87, 90, 93, 149, 159, 177–180, 181

L

Laplace transform, 127–128, 263, 264
Linear phase, 37, 56, 74
Long term evolution (LTE), 83, 169, 170, 175, 176, 176t, 177, 178, 179, 180, 182–183, 186, 188, 237, 238, 253–254
Luminance, 223

M

Main lobe clutter, 206, 207, 209
Mantissa, 7–8, 7t, 254
Minimum distance, 130
Modulation, 73, 81–83, 94–95, 113, 114, 140, 142, 145–146, 148–150, 151, 161–162, 163–164, 166, 170, 173, 175–177, 180, 182–183, 185, 197
Multiple input and multiple outputs (MIMO), 180–181, 182, 248, 253–254

N

National Television System Committee (NTSC), 225
Numerically controlled oscillator (NCO), 27–28, 123

O

Open base station architecture initiative (OBSAI), 189–190
Orthogonal frequency division multiplexing (OFDM), 169, 170, 171, 173, 175–178, 181, 182, 183, 185, 185t

Orthogonality, 158, 163–164, 169, 171–173, 174, 182, 183

P

Peak to average ratio (PAR), 183–186, 188
Polar coordinates, 10, 11, 17–18
Power control, 162, 164–165
Progressive, 225, 226–227
Pseudo-random code, 151
Pulse compression, 196, 197, 208–209, 214
Pulse repetition frequency (PRF), 197–200, 203–204, 205, 206, 207, 208–211, 213–214, 216, 217, 218, 219, 220–221
Pulse shaping, 85–88, 90, 91, 93, 162

Q

Quadrature amplitude modulation (QAM), 82, 83, 83t, 87, 93–95, 114, 130, 145, 149, 166, 173, 175, 176, 186–187
Quadrature phase shift keying (QPSK), 81–82, 82t, 83, 83t, 84, 87, 93, 94–95, 114, 116, 149, 159, 160–161, 164, 173, 175–176, 182–183
Quantization, 7–8, 26, 27–30, 74, 111, 119–120, 125–126, 147

R

Radar range equation, 195
Radio detection and ranging (RADAR), 191–194, 195, 196, 197, 198, 199–200, 201–202, 203–204, 205, 206–209, 210–211, 213–215, 216–217, 219, 220–222, 238, 245, 253–254
Raised cosine, 88–95
RAKE receiver, 159, 177–178, 181
Range ambiguities, 198, 206, 213–214
Rayleigh fading, 148, 180
Red/Green/Blue (RGB), 223–224, 228, 231
Remote radio head (RRH), 189–190
Roll-off, 88, 88t, 89, 90–91

S

Shannon capacity, 142
Side lobe clutter, 206, 207
Signal-to-noise power ratio (SNR), 29, 119–120, 121, 142, 181, 182–183, 214, 215
Soft decision decoding, 139, 140–141
Soft handoff, 163, 166
Synthetic array radar (SAR), 213–214, 215, 216, 217, 218–222

T

Time division multiple access (TDMA), 94, 147, 148–150, 153, 158–159, 161–162, 163, 166–167, 169, 177–178, 181, 182–183, 185*t*

V

Variable rate vocoder, 162
Verilog, 246
Very long instruction word (VLIW), 235
VHDL, 246
Viterbi decoding, 125, 131, 132, 134–139, 140, 141, 147
Vocoder, 147, 149, 162, 166, 240–241

W

Walsh codes, 152–155, 155–159, 160–162, 163–164, 166, 167, 171
Weave, 225, 226–227
WiMax, 83, 93–94, 169, 185*t*, 188

Y

YCrCb, 223–224, 227*t*, 228, 231

Z

Z transform, 41, 75–76, 77–78, 267, 268, 269–270

Printed in the United States
By Bookmasters